ASTRONOMY
ALL THAT MATTERS

ASTRONOMY

Percy Seymour

First published in Great Britain in 2014 by Hodder and Stoughton. An Hachette UK company.

First published in US in 2014 by The McGraw-Hill Companies, Inc.

This edition published in 2014 by Hodder and Stoughton

British Library Cataloguing in Publication Data: a catalogue record for this title is available from the British Library.

Library of Congress Catalog Card Number: on file.

Paperback ISBN 978 1 471 80162 4

eBook ISBN 978 1 471 89575 2

10 9 8 7 6 5 4 3 2 1

Typeset by Cenveo Publisher Services.

Printed and bound in Great Britain by CPI Group (UK) Ltd., Croydon CR0 4YY.

Hodder and Stoughton policy is to use papers that are natural, renewable and recyclable products and made from wood grown in sustainable forests. The logging and manufacturing processes are expected to conform to the environmental regulations of the country of origin.

Hodder and Stoughton

338 Euston Road

London NW1 3BH

www.hodder.co.uk

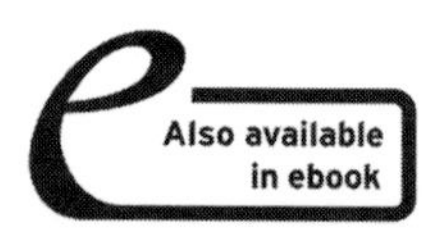

Contents

Introduction

On a clear moonless night, away from the lights of towns or cities, the Milky Way is the most obvious feature of the night sky. All the stars we see in the sky belong to the galaxy we know as the Milky Way; the nearby, brighter stars are spread more or less uniformly across the whole sky, but what we normally call the Milky Way is a shimmering ring of millions of much fainter stars forming a diamond-studded tiara arching over 'that inverted bowl we call the sky'[1]. We can picture the Milky Way as a disc-like city composed of 100,000,000,000 stars. Our Sun is one of these stars, about two-thirds out from the centre, along one of the radii of the disc. Most of the bright stars lie along spiral 'roads' leading out from the central concentration of stars.

Circling our Sun there are eight large bodies, the planets (including our own Earth), and a very large number of smaller bodies. Some of the planets have natural satellites, or moons, orbiting them. The distances between the stars of the Milky Way are more than 5,000 times greater than the size of our Solar System.

We now know that there are many thousands of galaxies strewn across the universe, and that they form the basic large-scale entities of the cosmos.

In this book we will discuss how our knowledge of the universe was gradually built up from the earliest beginnings of astronomy in the ancient world, right up to what we currently know about the extra-terrestrial universe. Crucially, we assume no previous knowledge of physics or astronomy. However, the essential concepts of these subjects – needed to understand the subject matter – will be introduced and explained as we go along.

We will see how the largely unbroken record of the movements of the Sun, Moon, planets and stars led to the realization that there were complex patterns in these motions, and these patterns led to the discovery of the basic scientific laws that govern celestial movements. In the 20th century we used these laws to help send men to the Moon and space probes to the distant reaches of the Solar System. These probes have considerably extended our detailed knowledge of the planets, their satellites and the minor bodies of the Solar System.

A multiplicity of new types of telescope, and the use of special instruments carried on satellites, has considerably extended our understanding of stars, galaxies and the universe, but this increased knowledge has also posed new problems for the basic physics that we use to interpret our observations. All these different topics will be discussed as we journey from our Earth to the edges of the observable universe.

Landing a man on the Moon

I believe that this nation should commit itself to achieving the goal, before this decade is out, of landing a man on the Moon and returning him safely to the Earth.[2]

President John F. Kennedy

US President John F. Kennedy's important speech of 25 May 1961 became a commitment that was to have a profound effect on science and technology, since it would also be the springboard for investigations into planetary and space science and it would also open up new areas of research for astronomy.

In order to achieve this goal there were a few scientific facts that needed to be established about our nearest neighbour in space.

First, how far is the Moon from Earth? And how big is it? It would obviously take a few days to get to the Moon and, since it is a moving target, scientists had to have something like the equivalent of a lunar bus timetable of the Moon's motion. Since any spacecraft going to the Moon had to move in the gravitational fields of the Earth, Sun and Moon, scientists had to consider the law of gravitation and the laws of motion – the basic 'traffic rules' of space. Additionally, scientists needed a three-dimensional map of the lunar surface and, adding layers of difficulty, this detailed mapping had to be compiled from the surface of the Earth.

Somewhat fortuitously, astronomy is the oldest of the exact sciences, so the accumulation of the data needed to send a man to the Moon really started in the ancient world.

The Moon as a calendar

The need of early hunter-gatherer communities for a calendar set in motion the science of astronomy. These people noted the links between the availability of food in the form of berries on trees, birds in the sky, animals roaming the countryside and the celestial cycles of the stars, the Moon and the Sun. However, with the coming of settled agricultural communities, knowing how many days there were in an annual cycle became an important goal for the early watchers of the skies. This was not an easy task, and the required knowledge had to be accumulated over several centuries.

Many ancient societies used the lunar (or Moon) monthly cycle as the basis of their calendars. The length of the lunar month is very nearly 29½ days. The seasons of the year form part of the solar year, which we now know is very nearly 365¼ days long. A simple calculation shows that 12 lunar months are therefore equal to 354 days, which is 11 days short of one year. So, extra days were inserted at the end of the year in ancient calendars to bring the lunar-based calendar back in step with the seasons.

Initially, the ancient Babylonians started their lunar month with the actual day on which the thin crescent Moon was first sighted in the west, just after sunset. Sometimes, because of adverse weather conditions, this first sighting was missed, so the start of the month had to be delayed. To end the necessity of depending on an actual sighting, the Babylonian astronomers invented numerical procedures that enabled them to predict, in principle, when they should be able to see the crescent Moon. As far as we are aware, this was the first time in the history of science that arithmetical and numerical procedures were used to generate an almanac, which could be used to anticipate actual sightings of a celestial object.

Eclipses of the Sun and Moon

A **total eclipse** of the Sun occurs when our Moon moves in front of the Sun, thus completely blocking the actual disc of the Sun from particular locations on Earth. This phenomenon provided the first clues concerning the relative sizes of the Moon and the Sun, as well as their relative distances from Earth. Greek astronomer Aristarchus of Samos (c. 310 BC–c. 230 BC) was the first to use an eclipse to make estimates of the Sun–Earth and Moon–Earth distances. Since the Moon is able to cover the Sun completely during a total solar eclipse, it implies that the ratio of the diameter of the Sun to that of the diameter of the Moon must be the same as the ratio of their distances from Earth.

A **total lunar eclipse** occurs when the Moon passes into the shadow of Earth. By timing how long the Moon spent in the shadow of Earth, Aristarchus was able to gain yet more data relevant to his calculations.

Since the Moon has no light of its own, only half of its surface will be lit by the Sun. Aristarchus reasoned that when the Moon was at first quarter or last quarter (when only half of the Moon's face, as seen from Earth is lit up), then the angle between the Sun and Earth, as seen from the Moon, must be a right angle. By measuring the angle between the Sun and Moon at such a time, one could use the properties of right-angled triangles to find the ratio of Earth–Moon distance to that of the Earth–Sun distance.

Hipparchus and the distance to the Moon

The methods used by Aristarchus of Samos were not sufficiently accurate to give a reasonable measurement for determining the distance to the Moon. The Ancient Greek astronomer Hipparchus (c. 190 BC–c.120 BC) used the method of **parallax** to find the distance. The basic principle of parallax can be easily demonstrated with a simple experiment. Hold out your hand at arms-length with your thumb upright. Look at the position of your thumb, against the background objects, first with your left eye and then with your right eye. Even if you keep quite still, your thumb will seem to change its position against the background. This phenomenon is called parallax.

The Moon is our nearest neighbour in space, so it will change its position against the background of stars, or with respect to the Sun, if observed from two different places on the surface of the Earth. Hipparchus observed a total eclipse of the Sun at Syene (now Aswan) and got a fellow astronomer to observe this eclipse at Alexandria. Whereas the whole of the Sun was covered at Syene, only one fifth of the disc of the Sun was covered at Alexandria. This was because parallax had caused the Moon 'to move' with respect to the Sun. Using this information Hipparchus calculated the distance to the Moon.

Navigating by the stars, Sun and Moon

For hundreds of years seamen used the stars for plotting their direction. The Sun and stars could also be used to find **latitude**. Latitude is the distance measured north or south of the equator of the Earth; that is, the North Pole is 90° north of the equator and the South Pole is 90° south of the equator. In the Northern Hemisphere, the height of the Pole Star (which is almost directly overhead at the North Pole) above the horizon can be used to give a navigator a very good idea of his latitude.

The Sun could also be used to find latitude. On one of the days of the **equinox**, falling either on the 20/21 March and 22/23 September, when we have equal day and night all over the world, then the Sun will be directly overhead somewhere along the equator. If a ship's navigator were to determine the altitude of the Sun, using a sextant, at local noon, then he would be able to determine his latitude. This method is true for the Northern and Southern hemispheres.

Finding **longitude**, how far east or west we are of some reference meridian (a meridian being a semi-circle starting at the North Pole, crossing the equator at a right angle, and ending at the South Pole), was a much more difficult task. Several ships and many lives were lost because seamen did not know their exact longitude. Longitude is connected with time. The Earth spins on its own axis, through 360°, in 24 hours, which means it moves through 15° every hour. Two places on the Earth's surface separated by 15° of longitude will be separated by one hour of time. If one could compare the time at two places on the surface of the Earth, then one could determine the longitude difference between these places. Local time at sea could be found by the Sun or the stars, but how does one know the time at some other place to make the comparison? Before the invention of the timekeeper known as the **chronometer**, and the radio, this task seemed impossible. Then Le Sieur de St Pierre, a Frenchman, proposed that the Moon could be used as clock.

As the Moon moves around the Earth, it moves through about 13° per day, on average, against the background stars. This means that the Moon is like the hand of a clock and the stars are the face of the clock. If the Moon was carefully studied from one observatory on the surface of the Earth, one could then work out where the Moon would be, at some future date, as seen from this observatory. The Moon's position against the background stars could then be published in a set of tables, which came to be known as the Nautical Almanac. The navigator of a ship, armed with such an almanac, could work out what the time was at the observatory where the observations were made by using an instrument called a **sextant** to measure the position of the Moon in the sky. He could find his own local time using the stars, and the longitude difference between his position at sea and that of the observatory could then be worked out. King Charles II established the Royal Observatory in Greenwich Park in 1675 to carry out the necessary observations that would eventually lead to the production of the first Nautical Almanac.

The Royal Observatory and Isaac Newton

The first Astronomer Royal, the Reverend John Flamsteed (1646–1719), made it his principal task to measure, as accurately as possible, the position of the stars to produce the best star catalogue and maps for that time. Flamsteed's work thus defined the face of the lunar clock. The second Astronomer Royal, Edmond Halley (1656–1742), made a very careful study of the Moon's motion. It soon became evident that searching for patterns in the Moon's motion, and trying to extend these into the future, was not an accurate enough way of generating the Nautical Almanac. What was required was a theoretical framework for calculating future positions of the Moon. This framework was provided by Newton's laws of motion and his law of gravitation.

Newton's work was far more radical than that of any of his predecessors, because his laws were universal. In other words, they could be applied to all large-scale phenomena anywhere in the universe. These laws made it possible to discover new physical laws, which could be useful to terrestrial science, in an astronomical context, and Newton's own work on motion and gravitation provided us with the first examples.

The **first law of motion** states that: A body continues in a state of rest or uniform motion in a straight line unless it is acted on by a force that would tend to change that state of rest or uniform motion.

The **second law of motion** says that: The rate at which a body changes its speed, if it is moving in a straight line, or the rate at which it changes direction, if it is moving at a constant speed, depends on the strength of the force exerted on it and on its mass.

The **third law of motion** asserts that: If one body exerts a force on another body the second body will exert a force on the first that is equal in strength but in the opposite direction.

Newton's **universal law of gravitation**, paraphrased into plain English, states that every particle in the universe attracts every other particle and that this force of attraction is proportional to the masses of both bodies, and it gets much weaker as the distance between the bodies increases.

The law of gravitation is not in itself sufficient to explain planetary motion; it has to be combined with the laws of motion. According to one of these laws, bodies tend to move in a straight line unless they have forces acting on them. We are all familiar with the consequences of this law when we drive around a corner in a car. The car and our bodies really wish to continue in a straight line, but the forces of the road on the wheels, and the wheels on the car, impel the car to change direction when we turn the steering wheel. A planet in the Solar System, like any other body, has a tendency to move in a straight line. However, the force of gravitation acting between a planet and the Sun, say, causes the planet to fall towards the Sun... we say it 'orbits' the Sun. Similarly, the Moon and all artificial satellites are in essence falling towards the Earth. However, orbiting bodies do not actually fall and hit the surface of the body being orbited because the body being orbited, that is the Sun or the Earth, is spherical and its surface is constantly curving away from the orbiting body.

Newton's laws of motion and his law of gravitation are thus able to explain the movements of the planets around the Sun and the movements of moons around their planets. Their impact on astronomy was tremendous and they still continue to play a major role in many of our efforts to understand the structure of the universe.

Charting the surface of the Moon

A final set of data was needed before men could be landed safely on the Moon. This was a set of cartographic charts to help land the lunar module. Although such charts existed they were not sufficiently accurate for the purpose. Professor Zdeněk Kopal suggested that the best way to make relief maps of the lunar surface with a higher level of accuracy was the **cinematographic method**. Just after sunrise, as the Sun gets higher in the sky, the lengths of the shadows that it casts of trees, telegraph poles and buildings decreases with the increasing height of the Sun. However the changes in the lengths of these shadows will not be uniform, because the ground on which the shadows fall will seldom be completely flat and level. The variations in the lengths of the shadows can be used to work out the gradients of the land. Towards sunset the lengths of the shadows will increase in a similar way. Professor Kopal suggested that topographic maps of the Moon could be produced by taking dozens of photographs of the Moon's surface from the Earth using a very powerful telescope in a variety of different relative positions of the Sun-Earth-Moon.

> *'And so we embarked on our work, originally intended only to explore the full capabilities of the cinematographic method as a tool for three-dimensional mapping of the Moon, but its scope did not remain so limited for long...*
>
> *...In response to the increased urgency of the task which we had initially agreed to undertake as a piece of research, and in recognition of the promise borne out by it, early in 1960, we [The Astronomy Department of Manchester University] were "upgraded" to a task force to keep pace with the other phases of the grand design which culminated in the Apollo landings of 1969 onwards.'*[3]

Professor Zdeněk Kopal in his book *Of Stars and Men*.

First the sponsorship of the project was extended from Cambridge [USA] Air Force Research Laboratories to include the Force's Aeronautical Chart and Information Centre, which had been commissioned by the National Aeronautics and Space Administration (NASA) to provide the maps for all lunar landings.

The photographs for the project were taken at the Pic-du-Midi Observatory, in France, high in the Pyrenees, and the photographs were developed in Manchester and in the US, and there was a joint collaboration between Manchester and NASA to reduce these plates, so that maps could be produced for the lunar landing.

The Apollo Programme

▲ **Apollo 11 lunar module pilot Buzz Aldrin, photographed on the Moon by mission commander Neil Armstrong, July 1969.**

The Apollo Programme was the one chosen to fulfil President Kennedy's commitment to land a man on the Moon and return him safely to the Earth.

In total, there were 17 Apollo missions. The earlier missions were used to test the various components – the Saturn launch rockets, the command and service modules and the lunar module – for the actual first landing, which ultimately became Apollo 11.

The method selected for the flight programme was lunar orbit rendezvous. The astronauts were to travel to and from the Moon in the command module, which contained the controls and the instrumentation. Rocket engines and fuel supplies were housed in a separate service module. On entering lunar orbit, the command module pilot would remain in the command module, while the commander and the lunar module pilot made the landing on the Moon in the lunar module. On completion of the mission on the Moon's surface, the descent stage of the lunar module was to remain on the Moon, and the ascent stage was to carry the astronauts back into orbit to rendezvous with the command and service module. The craft then embarked on its return journey, the lunar module being jettisoned. The service module was jettisoned just prior to re-entering the Earth's atmosphere.

The first man to set foot on the Moon was, of course, Neil Armstrong who, on 21 July 1969, uttered the now famous words:

> *'That's one small step for [a] man, one giant leap for mankind.'*[4]

There has recently been some controversy about the inclusion of the word 'a', but this is what Armstrong believed he said, and that is good enough for me.

What we learned from the Apollo missions

NASA has listed the top ten scientific discoveries made during the Apollo exploration of the Moon:

1. A study of the lunar surface and the lunar rock returned to Earth by the astronauts shows us that the composition of the Moon is very similar to that of the **terrestrial planets**. The terrestrial planets are those that have a similar structure and composition to our Earth – Mercury, Venus and Mars.
2. The ancient Moon still preserves its early history, a legacy that must be common to all the terrestrial planets. The extensive record of meteor craters on the Moon provides a key for unravelling timescales for the geological evolution of Mercury, Venus and Mars, based on their individual cratering records. Since the Moon has no atmosphere its surface has not been disturbed by winds or water erosion, and this is why its history has been so well preserved, unlike that of the other planets.
3. The youngest Moon rocks are virtually as old as the oldest Earth rocks. The earliest processes and events that properly affected both bodies can now be found only on the Moon, because of the geological processes on Earth, such as wind and water erosion, which have altered the surface.
4. The Moon and the Earth are formed from different portions of the common reservoir of materials. The

distinctly similar oxygen compositions of Moon rock and Earth rock clearly show a common ancestry. Relative to the Earth, however, the Moon was highly depleted in iron and in volatile elements that are needed to form atmospheric gases and water.

5 The Moon is lifeless. It contains no living organisms, fossils or native organic compounds.

6 All the lunar rocks originated through high-temperature processes with little or no involvement with water. The important discovery was that early in its history the Moon was melted to great depths to form a 'magma ocean'. The lunar highlands contain the remnants of early, low-density rocks that had floated to the surface of the magma ocean.

7 The lunar magma ocean was followed by a series of huge asteroid impacts that created basins that were later filled by lava flows.

8 The Moon is slightly asymmetric in bulk form, possibly as a consequence of its evolution under Earth's gravitational field. Its crust is thicker on the far side, while most volcanic basins – and unusual mass concentrations – occur on the nearside.

9 The surface of the Moon is covered by a rubble pile of rocky fragments and dust, which is called the **lunar regolith**, and which contains a unique radiation history of the Sun, which is of importance to understanding climate change on Earth.

10 The Apollo programme also had an enormous effect on technology. Glynn Lunney, who was once a flight director with NASA, said:

> *'Apollo really did drive our industry. We were asking people to do things that were probably 10 to 20 years faster than they otherwise would have done. And they knew it. They stepped up to it and succeeded. Today's cell phones, wireless equipment, iPads and so on are a result of the fact that the country did this hi-tech thing and created this large portfolio of available technologies.'*[5]

It was not only the electronics industry that benefited from Apollo; rocket and launching technology opened the door to using satellites for a large variety of different purposes. It also provided the staging ground for sending space probes to the planets (see Chapter 3).

Further important contributions to space exploration

First Soviet steps in space

The American impetus to enter the 'space race' was largely triggered by the first major Soviet Union effort. On 4 October 1957 the Soviets launched Sputnik into orbit around the Earth. This was the first artificial moon, or satellite, ever put into space, and it was this event that galvanized the US into action.

The Soviets also achieved a few important firsts as far as the Moon was concerned. On 4 February 1959 they launched the probe Luna, which made a flyby of the Moon on 6 February. It took the first pictures of the far side of the Moon. Since, very nearly, the Moon always keeps the same face directed towards the Earth, nobody had until this point seen the far side of the Moon.

Another probe, Luna 9, was a soft lander on the Moon and took the first photographs from its surface. Although the Soviets never sent a man to the Moon, some of their probes made important contributions to our understanding of our nearest natural neighbour in space.

The Chinese venture to the Moon

On 1 December 2013 the Chinese sent a probe to the Moon; on 14 December the Chang'e 3 mission landed on the Moon. This was the first time in 40 years that a space probe had touched the Moon's surface. Chang'e is named after the goddess of the Moon in Chinese mythology, and the rover that it took to the lunar surface, Yutu, is named after the goddess's pet jade rabbit. The rover carried an array of cameras and scientific equipment for investigating the geology of the Moon.

Exploring the Solar System from our Earth

We see it as Columbus saw America from the shores of Spain. Its movements have been felt trembling along the far-reaching line of our analysis, with a certainty hardly inferior to that of ocular demonstration.[6]

Sir John Herschel in 1846, talking about the prediction that there is a 'new planet' as yet undiscovered

The Solar System is made up of our Sun and a large number of bodies that orbit it, as well as satellites (or moons) that orbit some of the planets.

The latter half of the 20th century was marked by the detailed exploration of the Solar System by space probes. Once again, before this could be done scientists had to know the distances to the planets, the speeds at which they moved and whether Newton's Laws continued to be valid in the distant reaches of our Solar System.

The ancient Greek astronomers were the first to ask the scientific question: 'How are the Sun, Moon, planets and stars arranged in space?' They constructed geometrical models, based on careful observations of the movements of the celestial bodies, to arrive at an answer. After these early attempts, and after many centuries, astronomers settled on Kepler's laws of planetary motion (see below). These provided us with a scale model of the Solar System, but did not give us knowledge of the precise distances to the planets. Furthermore, the methods used to find the distance to the Moon could not be used to find the scale of the Solar System because the planets are so much further away. So other methods had to be developed. The first useful method was to use the transit of Venus across the disc of the Sun, which was used to find the distance of the Earth from the Sun, and hence provide a scaling factor for the entire Solar System.

Astronomy and astrology in the Ancient World

The ancients used the Sun, Moon and stars for calendar-making, time-keeping and direction-finding purposes, but the very complex regularities of the planets could not be used for these purposes.

We derive the word 'planet' from the Greek 'planetes', which means 'a traveller' or 'a wanderer'. Whereas the stars kept their distances with respect to one another, and hence could be linked together to form constellations, the Sun, Moon, Mercury, Venus, Mars, Jupiter and Saturn constantly changed their positions against the background stars, and this was why they were called wanderers, or planets. Only after Nicolaus Copernicus (1473–1543) did his work were the five 'naked eye' planets seen as distinct from the Sun and Moon.

The astrological perspective

The principal reason why astronomers made a careful study of planetary motion is because of astrology. There arose, in places such as Assyria, Babylon, Egypt, Greece and the Roman Empire, the belief that events on Earth – the destiny of kings and queens, emperors and pharaohs, and the fates of empires – were somehow linked to the movements of the then-known planets. So the conviction grew that if one could predict planetary motion, then one could predict events on Earth and the course of the royal household. In order to do this, astronomers started to develop mathematical models to generate tables of the future positions of these bodies.

The Babylonians extended their lunar arithmetical techniques – described in Chapter 1 – to include the planets. However, the Greek astronomers resorted to geometrical models. The Greek philosopher Aristotle (384 BC–322 BC) believed that the Earth was a sphere in the centre of the universe, and this sphere was surrounded by a very large opaque sphere to which the stars were attached on the inside. The centre of this celestial sphere was the same as the centre of the Earth. Filling the space between these two spheres were seven transparent crystalline concentric spheres, each one slightly larger than the next, nesting one within the other, not dissimilar to Russian dolls. To each of the spheres was attached one of the planets. Starting from our Earth, the next one surrounding the Earth had the Moon attached to it, to the next one was attached Mercury, then we had the sphere of Venus, followed by that of the Sun, then Mars, then Jupiter and finally Saturn. However, this scheme could not explain the complexities of planetary motion, so geometrical modifications had to be introduced.

From Ptolemy to Copernicus

Claudius Ptolemy (c. AD 90–168), the Greco-Egyptian astronomer, was well aware of the problems of the scheme proposed by Aristotle, and because he wanted to make accurate planetary predictions for astrological purposes he introduced various mathematical devices to overcome these problems. He suggested that each planet moved about its own small circle called the **epicycle**, the centre of which orbited Earth in a larger circle called the **deferent**. However, even Ptolemy's scheme was not very accurate and eventually his work was completely overthrown by the work of Copernicus.

Copernicus's decision to revive and refine a Greek idea of a Sun-centred universe caused a revolution that had a profound effect on all aspects of Western thought. In his scheme, the Sun was at the centre of the Solar System and all the planets, including the Earth, revolved around the Sun. The Copernican theory was able to explain the phenomena of day and night, and the seasons, in terms of the Earth spinning on its own axis once in every 24 hours, and going around the Sun in about 365¼ days. The apparent movement of the Sun against the background stars was easily explained, since as the Earth went around the Sun, so from the Earth the Sun would appear to move against the background stars (always assuming we could see the Sun and stars at the same time!).

The Copernican theory was also able to explain, more naturally, why the planets sometimes seem to move in one direction and sometimes in the opposite direction against the background stars. The planets Mercury and Venus, being closer to the Sun than the Earth is, would take less time to go once around the Sun. This means that sometimes they will overtake the Earth on the inside (travelling between the Earth and Sun), and at these times they will be moving in one direction; at other times they will be moving to go around the back of the Sun, as seen from Earth, in the opposite direction. The outer planets – Mars, Jupiter and Saturn – move around the Sun further out than the Earth. At certain times the Earth will overtake these planets on the inside, and at these times the outer planets will seem to move in the opposite direction to their normal motion.

Despite his boldness in moving the Earth from the centre of the universe, Copernicus was still very reluctant to dismiss most of the other ideas of Aristotle. He stuck rigidly to the Aristotelian concept of circular motion, and so to describe the detailed motion of the planets he had to resort to the use of epicycles and deferents, although this time the deferents had the Sun as the centre, and the epicycles were much smaller than those of Ptolemy's scheme. However, he had paved the way for an even more radical rethinking of the basic ideas of ancient science and astronomy.

Tycho Brahe and Johannes Kepler

Tycho Brahe (1546–1601) was a Danish nobleman who made several very important discoveries in astronomy. One of his most important contributions was to produce a very accurate set of data on planetary motion. These data provided the basis on which Johannes Kepler (1571–1630) established his laws of planetary motion.

Ellipse properties

Central to these laws are the basic properties of the geometric figure called the ellipse, which is rather like a flattened circle. A line drawn from the centre of a circle to the circumference (the radius) is always of the same length, no matter in which direction it is drawn. With an ellipse, the length of a line drawn from its centre to the curve of the ellipse will depend on the direction in which it is drawn.

Suppose we have an ellipse with its long axis, called its major axis, running in an east-west direction. The line from the east (or west) to the centre will have a maximum length. A line drawn from the centre to the north (or south) will

have a minimum length. At two points along the major axis, equidistant from the centre, are the foci of the ellipse. These foci have an important property; if two lines are drawn from the same point on the curve of the ellipse to these foci, the sum of the lengths of these lines will always be the same, no matter from which point on the curve they are drawn.

His **first law of planetary motion** states that: All the planets go around the Sun in elliptical (rather than circular) orbits with the Sun at one of the foci used to represent a given planet's orbit.

His **second law of planetary motion** says that: A line drawn from a planet to the Sun will sweep out equal areas in equal lengths of times.

His **third law of planetary motion** asserts that: The square of a planet's time to go once around the Sun is directly proportional to the cube of its mean distance from the Sun.

The third law means that the planets closer to the Sun are going around the Sun faster than those further out. This is not only because the planets further out have larger orbits to complete, but also because they are travelling more slowly.

This law helps us to make a scale model of the Solar System. Since we can measure the periods of time they take to go once around the Sun, we can then work out their relative distances from the Sun.

However, in order to find the actual distances of the planets from the Sun (an important requirement for exploring the Solar System by means of space probes), we need to know the distance of our Earth from the Sun. Edmond Halley, the second Astronomer Royal, was the first to propose how this could be done.

Halley, the astronomical unit and the transit of Venus

Edmond Halley is best known for the discovery of the comet that now bears his name, but he was a very competent and versatile scientist in general and an outstanding astronomer. Here we are particularly interested in his suggestion concerning the transit of Venus.

The planets Mercury and Venus orbit the Sun closer than the Earth does, and so on occasion these planets can be seen moving across the visible disc of the Sun; this movement is called a **transit**. In 1716 Halley suggested a method of measuring the distance from the Earth to the Sun – a distance known as the **astronomical unit**. He postulated that the transit of Venus could be used to measure this unit by using something called the **parallax**, the shifting of position that comes from viewing an object from two different points on the surface of the Earth (see Chapter 1).

Consider two different astronomers, one at each pole, viewing the transit of Venus. The person at the North Pole sees Venus following one path across the Sun, whereas the person at the South Pole sees Venus following a slightly higher path, shifted a little to the North. Because we see the Sun as a disc these two different paths will have different lengths. Halley proposed that an easy way to measure the difference between these paths would be to time the transits, using the four phases of the transits: the first, second, third and fourth contacts. First contact is when Venus just touches the Sun, second contact is when Venus is wholly within the Sun, third conduct is when Venus is just about to leave the disc of the Sun, and fourth contact is when Venus begins to move away from the Sun. With these lengths known, the distance between the Earth and the Sun can very easily be calculated using trigonometry and Kepler's third law of planetary motion. Although Halley did not actually make use of the method, he nevertheless left instructions for other astronomers on how to use the method to calculate the astronomical unit.

In the 17th, 18th and 19th centuries a number of expeditions were organized to observe the transit of Venus, but many of these were beset with difficulties so very accurate measurements using this method were never made. The best measurements were not made until 1961, using radar astronomy.

In a radar astronomy experiment, radio signals are sent from a radio telescope on Earth to the planet Venus, and the telescope acts, in receiving mode, to pick up the radio echo from the planet. Since the speed of radio waves (which is the same as the speed of light) is known, one can work out the distance to Venus by measuring the difference in time of the transmission of the signal and its return to Earth. Once this was done we had a very good knowledge of the distances to all the planets, from the Sun, using Kepler's third law.

Exploring the Solar System with the telescope

Most people accept that Galileo (1564–1642) was the first astronomer to use a telescope to study celestial objects. In 1609, Galileo heard that a device consisting of two tubes with lenses at each end had been invented in Holland, and that with this instrument it was possible to get close-up views of distant objects. Within a few months of hearing this report, he had made his own telescope, and with it he made a number of discoveries that undermined the universe as conceptualized by Aristotle and the Ptolemaic spheres, and also strengthened belief in the Copernican scheme. Among

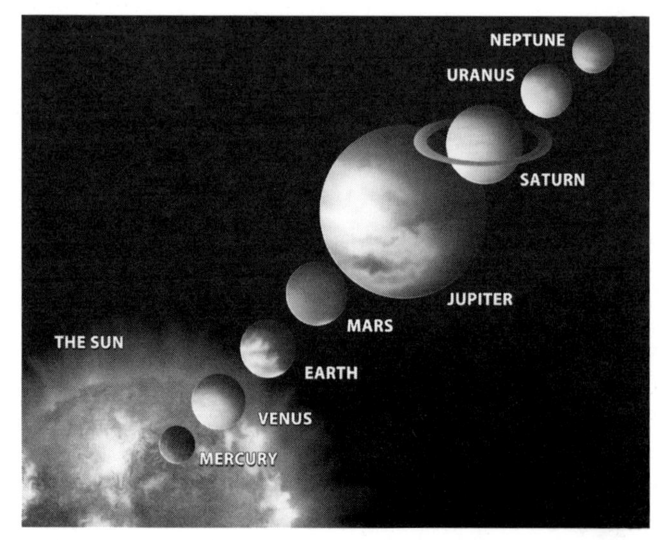

▲ The Sun and the eight planets of the Solar System.

the discoveries made by Galileo was that the surface of the Moon was covered with craters and mountains and that it resembled to some extent the surface of our Earth. He also discovered dark patches on the Sun, which we now call **sunspots**.

With this telescope he discovered that Jupiter had four moons circling around it, later named Ganymede, Callisto, Io and Europa. The planets have no light of their own; they only reflect light coming from the Sun, so it is only that part of the planet facing towards the Sun that is lit up. Galileo showed quite clearly that Venus underwent phases just as our Moon does, which also has no light of its own. This happens in the following way: Venus is orbiting the Sun closer than we are, and sometimes when Venus is very nearly between us and the Sun, far enough away from the Sun in the sky to be seen in the early morning or evening sky, it has its dark side facing towards the Earth, so it is seen just as a thin crescent of Venus lit by the Sun.

Also in the 17th century, other astronomers made telescope discoveries of the planets. The Dutch astronomer Christiaan Huygens (1629–95) discovered a satellite moving around Saturn that we now call Titan. Giovanni Domenico Cassini (1625–1712), an Italian-French astronomer working at the Paris Observatory, discovered a further four moons going around Saturn – Iapetus, Rhea, Tethys and Dione. He also discovered the rings of Saturn, including the gap in the ring system, dubbed the **Cassini Gap**.

Cassini, Rømer and the speed of light

Ole Rømer (1644–1710), a Danish astronomer working at the Paris Observatory, made an important physical discovery by studying the eclipsing of the satellite Io by its parent planet Jupiter. Rømer was working as one of Cassini's assistants when he decided to tackle a major problem that had surfaced concerning the innermost moon, Io. The observations made by Cassini seemed to indicate that the eclipsing of this satellite by Jupiter was not occurring at regular intervals, as he had expected. The times of the eclipsing of this moon by the planet were changing, in what appeared to be an unpredictable way. He ordered his assistants to do better observations and more precise calculations.

Rømer, however, did not believe that the observations or the calculations were at fault. He believed that the problem arose from the fact that the distances between Jupiter and Earth varied as they went around the Sun. Earth had a smaller orbit than Jupiter, and when the Earth was between Jupiter and the Sun the distance between the two planets was smaller than when Earth was about to go around the side of the Sun furthest from Jupiter. Cassini had predicted that Io would come out of a particular eclipse on 9 November 1676 at 5:25:45. However, when Io did emerge it was at 5:35:45. Christiaan Huygens used this data to come up with the first value for the speed of light, which he calculated to be 140,000 miles per second (225,000 km per second). This value is lower than the currently accepted value of 186,000 miles per second (300,000 km per second).

In the 20th century Albert Einstein (1879–1955) made the speed of light a cornerstone of his special theory of relativity.

The discovery of Uranus, Neptune and Pluto

The use of the telescope in astronomy led to the discovery of three planets.

The first planet to be discovered using a telescope was Uranus, which then became the seventh-known planet. It was discovered by William Herschel, on 13 March 1781, from the garden of his house in New King Street, in Bath, England. Initially he believed it to be a comet.

> *'The power I had on [he is talking about the magnifying power of his telescope] when I first saw the comet was 227. From experience I knew that the diameters of fixed stars are not proportionally magnified with higher powers, as planets are; therefore I now put on the powers of 460 and 932, and found that the diameter of the comet increased in proportion to the power, as it ought to be, on a supposition of its not being a fixed star, while the diameters of the stars to which I compared it were not increased in the same ratio. Moreover, the comet being magnified much beyond what its light would admit of, appeared hazy and ill-defined with these great powers, while the stars preserved that lustre and distinctness which from many thousand observations I knew they would retain.'*[7]

After a great deal of debate, this body was accepted as a planet. Herschel himself, in 1783, acknowledged this to the president of the Royal Society, Joseph Banks:

> *'By the observation of the most eminent Astronomers in Europe it appears that the new star, which I had the honour of pointing out to them in March 1781, is a Primary Planet of our Solar System.'*[8]

Herschel also discovered two of the five large moons of Uranus: Titania and Oberon.

Soon after the discovery of Uranus, people examining old records found that the planet had actually been spotted on 21 previous occasions, but because it was slow moving, most observers had catalogued it as a star. Astronomers tried to calculate the orbit of this new planet, but they soon ran into problems. The main difficulty was that it seemed impossible to reconcile the orbit calculated using the earlier observations with the orbit based on more recent observations. Two solutions were proposed to deal with this problem. The first was that Newton's law of gravitation broke down at great distances from the Sun; the second was that there was another planet beyond Uranus that was pulling it out of its predicted orbit.

The discoveries of Neptune and Pluto

Two mathematical astronomers set out to explore the second solution; that is, they tried to work out the orbit of the undiscovered planet that was changing the orbit of Uranus. The first astronomer to tackle the problem was a young mathematician from Cambridge called John Couch Adams (1819–92). In his first attempt he not only calculated the mass of the new planet but also predicted its position for 1 October 1845. He sent his predictions to Professor James Challis, the director of the Cambridge Observatory; he also sent his results to George Biddell Airy, the Astronomer Royal, and made two unsuccessful attempts to contact Airy.

In the same year François Arago, the director of the Paris Observatory, persuaded Urbain Le Verrier (1811–77) to start work on the problem of Uranus's orbit. Le Verrier took up the challenge immediately, but he had no idea that Adams was working on the same problem, and neither did Adams know that someone else was considering the issue. Once Le Verrier had calculated the orbit, he gave his results to the astronomers at the Paris Observatory, who made rather half-hearted searches for the new planet. He then sent his results to Johann Gottfried Galle (1812–1910) at the Berlin Observatory.

The following is an extract from an article on *History Topics: Mathematical Discovery of Planets*, written by J. J. O'Connor and E. F. Robertson on the MacTutor history of mathematics site http://www-history.mcs.st-andrews.ac.uk/HistTopics/Neptune_and_Pluto.html:

> *'Galle received the letter on 23 September [1846] and together with his assistant Heinrich d'Arrest began a search that night at the Royal Observatory in Berlin. D'Arrest suggested they use the latest star*

chart which had only just been produced. It took less than 30 minutes to locate a star not on their map. Of course they knew that they had found the "new planet" but they confirmed it the following night by observing it had moved relative to the stars.'[9]

Galle wrote to Le Verrier on 25 September 1846 saying: 'Monsieur, the planet of which you indicated the position really exists.' Le Verrier replied: 'I thank you for the alacrity with which you applied my instructions. We are thereby, thanks to you, definitely in possession of a new world.'[10]

This planet was Neptune.

The American astronomer Percival Lowell (1855–1916) had the idea that there was yet another planet, which was affecting the orbits of Uranus and Neptune. He set out to calculate the orbit of this planet, which he tentatively called 'planet X', and made observational searches for this body in the location in the sky, which it would have had on a particular date, but he never found it, although he searched for it from 1905 until his death. Thirteen years later the Lowell Observatory decided to restart the search for planet X. By now they had a more powerful telescope, built solely for this purpose. The observatory hired 23-year-old Clyde W. Tombaugh to use Lowell's predictions and the new telescope to search the skies for the planet. After a year of painstaking and detailed work Tombaugh did indeed find planet X. The discovery occurred on 18 February 1930 while Tombaugh was carefully examining a set of photographic plates created by the telescope. The Lowell Observatory was not ready to disclose the new discovery until more research could be done – the announcement came on 13 March 1930, on what would have been Percival Lowell's 75th birthday. This was Pluto!

Latterly, however, in 2006 the International Astronomical Union created a new definition for what constituted a planet and as Pluto did not meet all the criteria, it was downgraded from a planet to a dwarf planet.

Using comets to test Newton's Laws

In 1680 a bright comet appeared in the sky and Isaac Newton (1642–1727) and Edmond Halley watched it. Halley was very interested in comets, and decided to calculate its orbit. He discovered that the shape of its path was very close to the shapes of several others that had been seen in the past. Could this possibly be the path of the same comet? This would mean that the same comet was capable of returning on the same path time and again. The orbits were therefore elliptical and not parabolic. Whereas an ellipse is a mathematical curve that closes in on itself, a parabola does not close in on itself; it resembles, to some extent, a slightly opened out hairpin. Some astronomers believed that comets came from the depths of space, had their orbits bent by the Sun, and then returned to the outer reaches of the Solar System. The trouble was that at this time telescopes were so weak that only a small part of the path of comets could be tracked (when they came close to the Sun). In addition, the difference between an elliptical and a parabolic curve was so small that astronomers could not tell them apart from the visible small segment of their orbit. In 1682 another comet appeared. It was not as bright as that of 1680, but it was easily seen by the naked eye for some weeks.

Halley later made calculations relating to its passage and he assumed the orbit was a parabola, but again he saw similarities between this and others that had appeared in the past, namely those in 1607 and 1531. Perihelion distances (that is when it is closest to the Sun) were about the same, and the direction of motion in both cases was the same, in other words they were moving in the opposite direction to that of the Earth in its orbit around the Sun. Halley searched again and found that comets seen in 1455, 1378 and 1301 also had orbits similar to that of 1682. The interval in appearances seemed to work out at about 76 years. Halley predicted the return of the comet in 1758.

He knew that his predictions could not be exact. As it made its elliptical orbit around the Sun the comet would be affected by the gravitational pull of the large planets, Jupiter and Saturn, which, he felt, would alter the comet's speed and so make its return earlier or later than predicted.

In 1757 Alexis Claude Clairaut (1713–65), a brilliant French mathematician, decided that he would calculate the exact amount of influence that the planets in the Solar System would have on the comet, and so make a more precise estimate of its arrival time. He worked with two assistants for nearly six months until he had established that the comet would make its closest approach to the Sun on 13 April 1759.

In the meantime, Charles Messier (1730–1817), a French comet hunter, was scanning the sky for everything that might resemble a comet. On 21 January 1759, at about six in the evening, he observed a faint glow, which was the comet first noticed by Halley as the one that periodically returned to the neighbourhood of the Sun with a period of, roughly, 76 years, and whose precise orbit had been calculated by Clairaut and his assistants.

Messier's discovery was not the only one. Several European astronomers saw the comet soon afterwards. This proved to be a major triumph for Newton's laws of motion and his law of gravitation.

Professor Bernard Cohen, from Harvard University, a leading authority on Newtonian astronomy, in his paper 'Towards Newtonian Gravitation'; had this to say about Newton's interest in comets:

> *'That the theory of comets was a topic of major astronomical importance for Newton may be seen in the mass of observational data and computation he amassed concerning comets he actually saw and those reported in the literature. In all editions of* The Principia, *the theory of comets (not just their motions, but their composition) is the final scientific subject of the third and last book, occupying between a third and a half of that on the System of the World. No doubt the comets were of special importance in exhibiting the action of solar gravitational force to great distances, far beyond the limits of the visible solar system.'*[11]

Why was Mercury misbehaving?

We have already discussed that all the planets move around the Sun in elliptical orbits, with the Sun at one focus of each ellipse. If there were only one planet in the whole Solar System, then the major axis of its orbit would remain fixed with respect to the much more distant stars. However, the gravitational interaction between the planets causes the major axis of all the planets to move slowly with respect to the stars. The point of closest approach of a body to the Sun is an important point of reference, called the **perihelion.** However the gravitational attraction between the planets caused this point to move around the Sun. This is called the **perihelion precession.**

The perihelion precession of Mercury was more than expected on the basis of its gravitational interaction with the other planets. What was the cause of the small, but detectable, excess of Mercury's precession? Fresh from his triumph with the planet Neptune, Le Verrier proposed that there was another small planet between Mercury and the Sun. He worked out the orbit of this unknown planet, which was given the tentative name of Vulcan. Although a search was made for Vulcan, no convincing evidence for its existence was ever discovered. Two other suggestions were also proposed to explain this strange anomaly affecting Mercury. One was that the Sun was not exactly spherical and this would have an effect on Mercury, which was so close to the Sun. The other suggestion was that there was a ring of dust particles between Mercury and the Sun. No evidence was found for either of these theories.

The birth of Einstein's General Theory of Relativity

Some astronomers tried to modify Newton's law of gravitation, but all such attempts failed to yield convincing results, until Albert Einstein addressed the problem in 1915. Some ten years earlier, in The Special Theory of Relativity, Einstein modified Newton's laws of motion for bodies travelling close to the speed of light. One consequence of the special theory was that nothing can travel faster than light. Newtonian theory of gravitation required gravitational forces to work instantaneously across space. This violated Einstein's special theory of relativity and this was one reason why Einstein sought to modify Newton's law of gravity.

To understand the basic idea of general relativity, imagine a circular trampoline. It is perfectly flat. Now place a heavy metal sphere onto it and the trampoline will bend under the weight of the ball. The ball pushes the trampoline into a valley. The curvature of the material around the ball can be thought of as its gravitational field. If an object, such as a much smaller ball, enters the valley it will fall towards the bottom where the larger ball is situated. If the smaller ball moves sideways, then, when it enters the valley, it will move both sideways and towards the ball. This creates an orbit for the smaller ball, around the larger one.

The Sun can be seen as creating a valley in space and time, and one of the other objects in the valley, the Earth, does not roll towards the Sun because it is moving too fast. The curvature pulling the Earth towards the Sun is balanced by the Earth's own tendency to want to travel in a straight line, so it orbits the Sun. This is also how the Moon orbits the Earth. In the second case, Earth is the mass distorting the space-time around it and the Moon is the object orbiting it.

According to Einstein, matter tells space how to bend, and the curvature of space tells matter how to move.

General relativity has predicted a few things that are different to what Newtonian theory predicts. One of its most convincing tests came in 1919. Einstein's theory tells us that a beam of light just grazing the surface of the Sun will be bent by the curvature of space created by the Sun. In other words the curvature of space will allow us to see around the back of the Sun. Normally we wouldn't be able to see the stars around the back of the Sun, because the Sun is so bright. However, at the time of a total eclipse of the Sun, the Moon hides the Sun, and photographs taken of the area around the Sun during a total eclipse can then be compared with photographs taken earlier when the Sun was not in that part of the sky. This experiment was carried out in 1919 by two expeditions: one from Cambridge University headed by Professor Arthur Eddington and one from the Royal Observatory in Greenwich led by Frank Dyson. Later in 1919 they made an announcement to a joint meeting of the Royal Society and Royal Astronomical Society, at which they told the audience that Einstein's predictions were correct.

Soon afterwards Einstein was asked to write an article on the theory of relativity for *The Times*. In it he said:

> *'I respond with pleasure to your Correspondent's request that I should write something for* The Times *on the Theory of Relativity.*
>
> *After the lamentable breach in the former international relations existing among men of science, it is with joy and gratefulness that I accept this opportunity of communication with English astronomers and physicists. It was in accordance with the high and proud tradition of English science that English scientific men should have given of their time and labour, and that English institutions should have provided the material means, to test a theory that had been completed and published in the country of their enemies in the midst of war. Although investigation of the influence of the solar gravitation field on rays of light is a purely objective matter, I am none the less very glad to express my personal thanks to my English colleagues in this branch of science: for without their aid I should not have obtained proof of the most vital deduction from my theory.'*[12]

This bending of light by a massive object is called **gravitational lensing** because the gravitational field of an object, is, in effect, acting like a lens. We will discuss this matter again, in the final chapter, when we talk about the evidence for dark matter.

Comets, meteors, asteroids and other bodies

In this section we will briefly discuss some of the minor bodies of our Solar System, including comets, asteroids and meteors.

Comets

Essentially, comets are very large dirty snowballs, covered in a sooty-like material. They are mainly composed of water, carbon dioxide, ammonia, methane and dust. These materials are believed to have come from the time of the formation of the Solar System.

The dirty snowball is the nucleus of the comet. Because of their highly elliptical orbits, most comets spend a great deal of their time orbiting in the outer reaches of the Solar System. When they are moving towards perihelion – their closest approach to the Sun – the heat from the Sun melts some of the water ice to form an atmosphere around the nucleus. This extended atmosphere is called the **coma**.

Comets have two tails: a dust tail and a plasma tail. These are the parts of the comets we actually see since they reflect sunlight. Coming to us from the Sun is a stream of very energetic fragments of atoms that we call the **solar wind**. The dust particles are blown away from the Sun by the solar wind in a gentle curve, but the plasma tail consists mainly of fragments of atoms, and this tail is always blown by the solar wind directly away from the Sun. This means that when the comet is moving away from the Sun, the tail can actually be ahead of the coma. A dust tail curves more gently than the plasma tail, because it consists of dust particles that are heavier than atoms and ions.

Meteor showers and meteor streams

Meteor showers are displays associated with the Earth's passage through a meteor stream. These streams consist of debris left over from the passage of comets. As comets pass through the inner Solar System the radiation from the Sun causes them to heat up, evaporating the dusty icy materials of the comet. These particles are left in the wake of the comet's passage, creating a stream of small debris that is strewn along the orbit of the comet. If the orbit of the Earth intersects with the orbital path of the meteor stream, then at regular, predictable times throughout the year the Earth will pass through the stream of debris that creates a meteor shower.

The observed shower is caused by the tiny dust particles ploughing into the atmosphere of Earth at very high speeds. The friction they experience causes them to burn up, and what we see are the glowing remnants of this encounter. Some of the cometary fragments are large enough to survive this heat treatment, and if they actually land on the surface of Earth, we then call them **meteorites**.

Meteorites

Most of the larger meteorites come from the main asteroid belt, which lies between Mars and Jupiter. Although most asteroids (see below) are separated by thousands of kilometres there are, nevertheless, frequent collisions between them, and at these times fragments of the asteroids are knocked off, and sent into more elliptical orbits that may encounter our Earth's atmosphere. Occasionally, an entire asteroid can break apart but mostly a few small pieces are chipped off, which can travel through space for millions of years until they are caught by a planet's gravity.

The chemical analysis of meteorites has given space scientists a great deal of information about the gas and dust cloud out of which the Solar System was born.

Asteroids

The asteroids, which are sometimes referred to as minor planets, are a large number of mainly rocky bodies that orbit the Sun, mostly between the orbits of Mars and Jupiter. The region that they occupy in space is called the **asteroid belt**.

The first asteroid that was discovered was Ceres. It was spotted by Giuseppe Piazzi (1746–1826) in 1801. Vesta is the largest asteroid found so far, and was discovered by Heinrich Wilhelm Olbers (1758–1840) in 1807. These two bodies are now classed as dwarf planets, along with the former planet Pluto.

A French-Italian mathematician called Joseph-Louis Lagrange (1736-1813) worked out that two bodies can orbit the Sun, at the same distance from the Sun, provided that the two bodies and the Sun are at the three corners of an equilateral triangle.

Two groups of asteroids are to be found along the orbit of Jupiter, one group orbits ahead of Jupiter, called the

Trojan Asteroids, while the other group orbits behind Jupiter, and is called the Greeks. As seen from the Sun they make angles of 60° with Jupiter.

The Kuiper Belt

The Kuiper Belt is an elliptically shaped doughnut-like region in the outer reaches of the Solar System lying approximately 30 to 50 **astronomical units** from the Sun. (Distances in the Solar System are measured in astronomical units, one astronomical unit being the mean distance of the Earth from the Sun. In terms of miles one astronomical unit is about 93 million miles (150 million km).)

In some ways it is similar to the asteroid belt. However, whereas the asteroid belt is composed mainly of metal and rocky fragments, the Kuiper Belt objects are composed mainly of icy chunks of various substances. The objects in the Kuiper Belt are similar in composition to comets: a mixture of frozen water, ammonia and various hydrocarbons, such as methane.

The Oort Cloud

Still further out, by a very significant distance, is the Oort Cloud. This is a vast cloud of comets that stretches from about 50 to about 100,000 astronomical units from the Sun. This means that the comets can be gravitationally influenced, or **perturbed**, by the stars of our Milky Way galaxy. These perturbations can cause large changes in their orbits, and bring them closer in towards our Sun. Most of these comets will have very long periods.

Exploring the Solar System with radio waves

Radio astronomy and radar astronomy both developed out of the use of radar during the Second World War. Radar works in the following way. Suppose we bounce radio waves off a distant reflecting object, and we measure precisely the time delay between the sending out of the signal and the return of the echo, then we can work out the distance to the object. This is because radio waves travel at the same speed as light waves, and since we know this speed very accurately, the distance to the object is just half the time delay multiplied by the speed of radio waves.

While scientists were developing radar they made a few discoveries that are relevant to radio astronomy. One such discovery was that our Sun emitted radio waves. Another discovery was that the ionized gases in the vapour trails of meteors reflected radio waves and this fact could be used to find the distances to these trails. When radio astronomy was further developed, it was also discovered that the planet Jupiter was a source of radio waves.

The radio waves from Jupiter arise in the following way. Jupiter has a very strong and very large magnetic field. In other words, the planet behaves, to some extent, like an enormous bar magnet. Trapped in this magnetic field are charged particles, which are fragments of atoms. Charged particles in motion will spiral around the lines of force of a magnetic field, and when they do so they emit radio waves. These were the waves detected by radio astronomers. These observations showed that the radio waves were not coming from the planet itself, but from its extended magnetic field. The radio radiation from Jupiter seemed to 'wobble' in a strange way. Astronomers were able to explain this as being due to the inclination between the rotation axis of Jupiter and its magnetic axis. This allowed them to calculate the angle of inclination, which is about 11 degrees. This finding was eventually confirmed when space probes visited the planet.

Radar astronomy was used to make a better measurement of the distance to Venus, by timing how long it took radio waves to be sent to this planet and return to the Earth. Earlier in this chapter we saw that Kepler's laws of planetary motion gave us a scale model of the Solar System, but in order to find actual distances we needed to know the distance between two planets. Radar ranging of the distance to Venus provided us with this information. Venus has a thick cloudy atmosphere, so with optical telescopes astronomers were unable to measure the spin rate of the planet. Radio waves of the right wavelength were able to penetrate the cloud cover and measure the spin rate using the **Doppler Effect**.

In order to explain the Doppler Effect, we will use the analogy of the radar speed traps used by the police. The radar gun used by the speed traps sends out pulses of radio waves, at set time intervals, and if they are bounced off a stationary car the time intervals received back will be exactly the same as those sent out. If, however, the car is moving away from the radar gun, even in the short time interval between pulses, so the returning pulses will have a longer time interval between them. The change between the intervals of the outgoing beam, and that of the returning beam, can be used to find the speed of the car. As a car is moving towards the radar gun, the time intervals between pulses will be shortened. This shortening and lengthening of pulse intervals is called **Doppler Shifting**. If a radar beam is sent to Venus, with pulse intervals of a known length, the signals returned from the planet will not have the same spacing in time. This is because those reflected from the side of Venus moving towards the Earth will be different from those reflected from the side travelling away from Earth. This difference can be used to calculate the spin rate of the planet. These radar studies showed that Venus spins on its own axis once every 243 Earth days. We also know that it takes 225 Earth days for Venus to go once around the Sun. This means that a day on Venus lasts longer than its year!

Exploring the Solar System with space probes

I think we're living through the greatest age of discovery our civilization has known. We've voyaged to the farthest reaches of the Solar System. We've photographed strange new worlds, stood in unfamiliar landscapes, tasted alien air.[13]

Professor Brian Cox

The latter half of the 20th century was marked by the sending of space probes to the planets. These probes were fitted with a variety of laboratory-like instruments to make measurements in the near environment of the planets, and these measurements increased by a considerable margin our understanding of planetary environments. The probes also carried cameras that took amazing close-up photographs of the planets and their satellites. We learned more about the physical nature of the planets from the information sent back to Earth from these probes than we ever could have learned from the surface of Earth. But before we could successfully launch these probes on their voyages, we had to have a strategy to design the technology of the probes, and a plan for how we could launch probes from a moving Earth to planets that were themselves moving, without the expenditure of vast sums of money on propulsion. Early in the 20th century, long before we had in place the technology to send such probes, one man was thinking about the physics and astronomy of interplanetary flight. His name was Walter Hohmann (1880–1945).

Walter Hohmann – visionary of space travel

Walter Hohmann was the son of a doctor. He was born in Hardheim in Germany and lived for a short spell as a child in Port Elizabeth, South Africa, before returning to Germany to study at the Technical University of Munich, where he read engineering. While in southern Africa his father had showed him the southern constellations and the young Hohmann became interested in space; he also read the science-fiction works of the French author Jules Verne and the German author Kurd Lasswitz. While working as an engineer in Wrocław in Poland, Hohmann started to question how man could seriously engage in space flight.

In the 1920s, a space flight society, consisting of some of Germany's amateur rocket enthusiasts, was set up and Hohmann became a leading figure. The author Willy Ley asked Hohmann to write a paper on the possibility of space travel. Hohmann's paper was called (to give it its English title) 'Routes, Timetables and Landing Options'. In it he proposed using a separable landing module to land on the Moon. It was this idea that was subsequently adopted for the Apollo missions.

The best orbits for sending probes to the planets

In order to minimize the amount of fuel needed to send a probe from the Earth to a planet, Hohmann calculated what he called **least energy transfer orbits**. In such an orbit, a space probe becomes an artificial planet of the Solar System, obeying Kepler's laws of planetary motion as it journeys from our Earth to another body in the Solar System.

Consider what happens when we send a space probe to Mars. After leaving the Earth its path to Mars will be one half of an ellipse. The major axis of this ellipse will be the radius of the Earth's orbit around the Sun (we will assume this is very nearly circular) plus the radius of the orbit of Mars (also assumed to be circular). When it leaves Earth, it is at the perihelion (closest approach to the Sun) of its orbit around the Sun, and when it reaches Mars it will be at the aphelion (its furthest point from the Sun) of its orbit. However, if it is to take close-up pictures of Mars, the timing is, very naturally, crucial. Its launch from the Earth must be so timed that during its motion from perihelion to aphelion, Mars must be in the right position to encounter the probe.

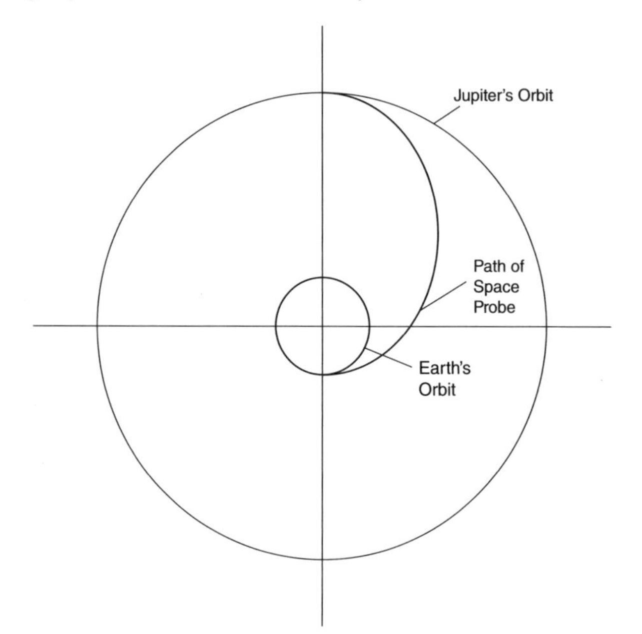

▲ **A typical example of a path of a planetary probe – Earth to Jupiter.**

Let us look at the case of sending a probe to one of the inner planets, such as Venus. When the probe is launched from Earth it will be at the aphelion of its elliptical path, and when it reaches Venus it will be at its perihelion point.

Once again the timing of the launch must be such that when it reaches its perihelion, Venus must also be close to this point, if it is to photograph and make measurements in the vicinity of the planet. In both cases the probes are moving under the force of gravitation, and obeying Newton's laws of motion, for the greater part of their orbit. Rocket fuel is only expended at the start, and at the end, either to do a fly-by of the planet or to make a soft landing on its surface.

For voyages to the outer planets, such as Jupiter, Saturn, Uranus and Neptune, similar orbits are used for the probe's path, or trajectory, but an additional technique is used. This is called **gravitational assist**. Let us look at how gravitational assist was used in the case of the space probe Voyager 2.

Voyager 2 was planned to make a grand tour of the Jovian planets Jupiter, Saturn, Uranus and Neptune. The spacecraft was launched on Hohmann transfer orbit to Jupiter and its arrival at Jupiter was carefully timed so that it would pass behind the planet. As the spacecraft came under Jupiter's gravitational influence, it fell towards the planet, increasing its speed towards maximum at its closest approach to Jupiter. Since all masses in the universe attract each other, Jupiter sped up the spacecraft substantially, and the spacecraft tugged gravitationally at Jupiter, causing the planet to lose some of its orbital energy. But, because of the great difference between the mass of the spacecraft and that of Jupiter, this loss of energy for the planet was very small.

The spacecraft passed by Jupiter, but since its velocity was great enough to overcome the tug of Jupiter's gravity it was not captured by the large planet. Of course it slowed down slightly relative to Jupiter as it climbed out of the planet's large gravitational field. However, relative to the Sun, it never slowed all the way to its initial Jupiter approach. It left the Jovian environment carrying an increase in momentum stolen from Jupiter. Jupiter's gravity served to connect the spacecraft to the planet's ample reservoir of momentum. In other words the gravitational link between Jupiter and the spacecraft transferred additional forward momentum to the craft. This technique was repeated at Saturn and Uranus.

In the rest of this chapter we will combine what we've learned about the planets from Earth-bound studies with what we've discovered using space probes to give an overview of our current knowledge of the main bodies of the Solar System.

The rotational axes and the orbital planes of the planets

All the planets spin on their own axes, and in this they act as free spinning gyroscopes. The angle between the axis of rotation of a planet and the plane of its orbit around the Sun varies considerably from one planet to the next. For Mercury this angle is 0° and for Jupiter it is about 3°. The inclination of Earth is about 23½°, for Mars it is just over 25°, for Saturn it is about 27° and for Neptune it is just over 28°. Venus is quoted as having an axial tilt of about 177°, which actually means it is very nearly at right angles to the plane of its orbit, but, uniquely, it is rotating in the opposite direction to the other planets. Uranus has the largest tilt of all the planets, with an inclination of 97°.

It is the 23½° tilt of Earth's axis that gives us the seasons of the year. In the northern winter the North Pole is leaning away from the Sun and in the northern summer it is leaning towards the Sun, giving us the longer days and warmer temperatures that we associate with summer. On around 21 March and 23 September, the poles of Earth are equidistant from the Sun, so we have equal days and nights all over the planet.

Basic properties of the terrestrial planets

The terrestrial planets are those that resemble the Earth in internal structure, viz Mercury, Venus and Mars.

Internal structure

We cannot see inside the planets, so we have to make mathematical models of their interiors, based on what we can see and measure, and then we have to combine this information with the known laws of physics. For those planets such as Earth and Mars that have natural satellites we can work out their masses from the orbital periods of their satellites. For Mercury and Venus we have to work out their masses from their mutual effects on their orbits. We also know the sizes of these planets, so we can work out their mean densities. For our Earth we have additional information about its interior from the study of how earthquakes move through the Earth.

For Earth and Venus we can identify four distinct regions. At the centre of each planet we have a solid inner core, surrounded by a semi-molten outer core, then the region called the mantle and finally a thin crust. Mercury has a semi-liquid inner core, which occupies a large part of its interior, surrounded by a mantle and then a thin crust. Mars has a small semi-liquid core surrounded by a large mantle and then the crust. Mercury was believed to be solid throughout. This was because its very small size meant it should have lost all the heat needed to keep some of its interior molten. However, space probes showed that it does have a small magnetic field. The generation of magnetic fields in planets is discussed later in this chapter.

Atmospheres

Mercury has a very thin atmosphere, and, very naturally, its pressure is also extremely low. It consists of hydrogen, helium, oxygen, sodium, calcium, potassium and water vapour. These components come mostly from the solar wind. The nature of its atmosphere was discovered in 1974 by the Mariner 10 probe, but improved measurements were made in 2008 by the Messenger spacecraft, which discovered magnesium in its atmosphere. The radiation pressure of sunlight pushes atmospheric gases away from the Sun, thus creating a comet-like tail. The atmosphere of Mercury is so tenuous and it is so close to the Sun that the temperature of Mercury is determined almost entirely by the Sun. During the day, temperatures can be as high as 450° C, hot enough to melt tin and lead. At night the temperature can be as cold as –170° C. Mercury has a greater range of temperature than any other planet.

The atmosphere of Venus is much denser than that of Earth. The temperature of the surface is 467° C, while its atmospheric pressure is very much higher than that of Earth. The atmosphere of Venus supports opaque clouds made of sulphuric acid; this is why Earth-based telescopes are unable to penetrate the cloud to see the surface. The main atmospheric gases are carbon dioxide and nitrogen with traces of other substances. The atmosphere is in a state of vigorous circulation and super-rotation. The whole atmosphere circles the planet in just four days, much faster than the planet's rotation speed of 243 days.

The Earth's atmosphere, when it is dry, contains 78 per cent nitrogen, 20 per cent oxygen and traces of argon and carbon dioxide. There are small amounts of other gases. It also contains variable amounts of water vapour, on average about 1 per cent. Although air temperature and atmospheric pressure vary at different levels, we know, of course, that they are suitable for the survival of terrestrial plant and animal life.

The atmosphere of Mars is very different from that of Earth in many ways. It is composed mainly of carbon dioxide, which makes up 95 per cent of the atmosphere, and it has much less nitrogen than on Earth, at approximately 3 per cent. It also has very little oxygen and water vapour.

Magnetic and particle environments

In 1974, Mariner 10 made the discovery that Mercury had a magnetic field, which, to an initial approximation, resembled that of our Earth, i.e. it was, to some extent, like a bar magnet. The strength of the field is only 1 per cent of the Earth's field, so it is not as effective as that of Earth in holding the solar wind particles at bay.

The discovery of the field came as a surprise, because the existence of planetary magnetic fields is usually explained in terms of what is called **dynamo theory**. According to this theory the movement of electrically conducting fluids in the interior generate electric currents, which give rise to a magnetic field, just as electric currents in the wire of an electro-magnet, generate its field.

It was generally believed that Mercury, because of its small size, had cooled over the years. There is another difficulty with dynamo theory, the fact that Mercury has a slow, 59-day rotation period that would have made it impossible to generate a magnetic field. These problems have still not been successfully solved. The flowing of the very fast solar wind interacts with the magnetic field of Mercury. It gives rise to what space scientists call a **bow shock-wave**, which is similar to the bow wave of a ship as it moves through the ocean.

The magnetic and particle environments of Venus were studied by Mariner and Venera space probes and also by the Pioneer Venus Orbiter. All these investigations showed that the strength of the magnetic field of Venus is less than 1/10,000th of the Earth's field. This result is somewhat surprising, since Venus probably has a liquid core and is comparable in size to Earth. The answer to the very low strength of its magnetic field may be due to its slow rotation. Another possibility is that no internal dynamo operates since it has a completely fluid core, which is divided into stable strata, rather like the layers in an onion, and as a result there is no adequate vertical movement to provide a source of energy to drive the convective currents needed to power the dynamo. However, currents induced in the conducting ionosphere of Venus by the solar wind prevent this solar wind from reaching its surface and as a result this planet also has a well-developed bow shock-wave.

Measurements of the magnetic field of Mars were made by Soviet Union and US space probes. The Soviets interpreted their results as showing that Mars has a small field, which is similar to that of a bar magnet, but the Americans interpreted their results as a magnetic field caused by the solar wind. At the end of the 20th century the Mars Global Surveyor probe made magnetic measurements at 250 miles (400 km) above the Martian surface. These measurements were used to produce the first global magnetic map of Mars. This map shows that intense magnetism is confined to the highly cratered highland regions in the south. This result seems to indicate that, at one time, Mars did have an internal dynamo, which stopped operating a long while since.

The first person to study Earth's magnetic field was William Gilbert (1544–1603), who was physician to Queen Elizabeth I. By making use of information on compass directions collected by English seamen on their voyages all over the world, Gilbert formulated a theory for the Earth's magnetic field. He was able to show that the magnetic field of a magnetized sphere was similar to that of Earth. In other words, he showed that the Earth behaved as if it had a bar magnet situated close to its centre, and almost aligned with its rotation axis. Modern work on the magnetic field seems to show that the major part of the Earth's field can be described by imagining a bar magnet, with its centre situated about 250 miles (400 km) from the Earth's centre, and with its axis aligned by 11.5° to Earth's rotation axis.

Satellites have been used for several decades to map the magnetic field of our Earth far above its surface. These satellite measurements have shown that the field is confined within a region, the **magnetosphere**, which is compressed on the sunward side and drawn out into a long tail on the opposite side. The electrically charged particles from the solar wind cannot cross the lines of force of the Earth's field, but stream past it making a bow shock-wave. Somewhere behind the Earth the various strands of the solar wind join up together, thus enclosing the Earth's field in a pear-shaped region, the **magnetotail**. Also trapped in the magnetosphere are the charged particles of the **Van Allen radiation belts**. The **aurora borealis** (the Northern Lights) and the **aurora australis** (the Southern Lights) are associated with the two Van Allen belts.

The aurorae can best be described as moving curtains of light. Mostly these curtains are bluish-green, but red aurorae are also seen on some occasions. The extensive use of sky cameras, jet planes and satellite observations has been used to determine the region in which the maximum number of aurorae occur. This region is called the **auroral oval**, and it is actually the intersection of the outer shell of the Van Allen radiation belt with the Earth's atmosphere. The auroral light is emitted by atoms and molecules of different gases, high in the upper atmosphere, which had been excited by collisions with the energetic particles present in the belt.

Basic properties of the Jovian planets

The planets Jupiter, Saturn, Uranus and Neptune have more in common with each other than they have with the terrestrial planets, and hence we call them the Jovian planets.

Internal structure

Jupiter and Saturn mainly consist of hydrogen and helium. The outer layers of these planets are composed of molecular hydrogen whereas the inner layers are liquid metallic hydrogen and their inner regions are composed of rocky and icy materials. Liquid metallic hydrogen is an unusual form of hydrogen, only recently reproduced in laboratories. The fact it is in liquid form means that if you pour it into a vessel it would assume the shape of the container, and not spread out throughout the entire volume as a gas would. The fact that the hydrogen is metallic means that it can conduct electricity. The high pressure of the material in this area of these two Jovian planets means that the electrons and protons together form a fluid (there is no atomic structure as such), and this is what makes them able to conduct electricity.

Mathematical modelling of the interior of Uranus leads us to believe that this planet has a rocky interior made of silicates, an icy mantle or perhaps an ice-rock mixture, made up of water ice, methane ice and ammonia ice. Modelling of the interior of Neptune suggests that it is very similar to Uranus.

Atmospheres

The atmospheres of Jupiter and Saturn are composed of different types of clouds: ammonia clouds, ammonium hydrosulfide clouds and water clouds. Uranus and Neptune have mainly methane clouds.

Magnetosphere

All the Jovian planets have very extensive magnetic fields, but the angle between the rotation axis and the magnetic axis of each planet varies a great deal. For Jupiter this angle is 9.6°, for Saturn 0°, for Uranus it is 59° and for Neptune it is 47°.

Spaces probes have also photographed aurorae in the atmosphere of Jupiter.

Planetary rings

As seen from Earth, using telescopes, Saturn was for a time the only planet known to have rings. In 1977, using terrestrially based telescopes, and an ingenious method, a ring system was discovered around Uranus. The method, one often used to study the sizes and atmospheres of the planets, is called the **method of occultation**.

A planet is an opaque body, even if the outer planets are largely gaseous, and we only see it because it reflects sunlight. However, when a planet passes in front of a star, it will blot out the light from the star. Such observations were made using Uranus as the occulting body. As Uranus started to move in front of the star, but sometimes before it actually reached it, the light from the star seemed to be twinkling with reduced light. The same thing happened again on the other side, as Uranus started to move away from the star. Astronomers interpreted these observations as indicating that Uranus had a ring system. In 1979 Voyager I discovered a small ring system around Jupiter, and in 1989, Voyager 2 discovered a ring system around Neptune. All these rings are made up of small bodies orbiting these Jovian planets. Space probes took photographs and made measurements of these systems of rings.

Satellites of the planets

Of the eight major planets of the Solar System, only Mercury and Venus do not have moons, or natural satellites. If the two innermost planets ever had moons, their orbits would have been affected by the Sun, and they would have lost them to the Sun a long time ago. We have already discussed our own Moon in the first chapter, so now we move on to the moons of the other planets.

Mars has two moons, called Phobos and Deimos; they are believed to be captured asteroids. Phobos is 17 miles (27 km) long in its longest dimension, and Deimos is 9 miles (15 km) long. Both are cratered and orbit the planet in rather low orbits; Phobos is only 3,000 miles (4,800 km) above the Martian surface and orbits in little over seven hours. Deimos is further out and orbits in about 30 hours.

Jupiter has the largest number of moons of all the bodies in the Solar System – 67 in total. The four largest moons – called the Galilean moons, after their discoverer Galileo – are Io, Europa, Ganymede and Callisto. The

Voyager Spacecraft took photographs of these satellites, which showed that on the surface they are very different to one other. Io is the most volcanically active body in our Solar System. At least three of its volcanoes were erupting while the photographing was in progress, but there was also substantial evidence of volcanic craters and lava flows. Europa's surface and crust is made, almost entirely, of water ice. Ganymede's surface is also made of water ice. Callisto is one of the largest and most heavily cratered of all the planetary satellites.

Saturn has 62 moons, the largest of which is Titan. After Jupiter's moon Ganymede, Titan is the largest of all the moons. It is the only natural satellite known to have a dense atmosphere, and there is also evidence for stable bodies of surface liquid.

Uranus has 27 moons, of which the largest are Titania and Oberon. Neptune has 14 moons, of which the largest is Triton.

Conclusion

In this chapter we have looked at how the exploration of the Solar System, with space probes, has considerably expanded our knowledge of our near environment in space. Over more than 2,000 years this system has provided a model of how quantitative exact science should be done. It has inspired a great deal of outstanding work in mathematics, physics, astronomy and technology, and in the 20th century it provided the inspiration for our first model of the atom. In the next chapter we will see how our understanding of the properties of light has allowed us to probe much deeper into the universe.

Decoding the message of starlight

That was Bohr's marvellous idea. The inside of the atom is invisible, but there is a window in it, a stained-glass window: the spectrum of the atom. Each element has its own spectrum, which is not continuous like that which Newton got from white light, but has a number of bright lines which characterize that element.[14]

Jacob Bronowski

Codes and ciphers have played significant roles at several stages of human history, but never more so than during the Second World War. On 4 and 5 February 2014, there were numerous press articles highlighting the importance that code-breaking played in shortening the war. In particular, they drew attention to the fact that 5 February was the 70th anniversary of the first use of a machine, called Colossus, to break a German encoded message. Writing in *The Daily Telegraph* on 5 February 2014, Sophie Curtis said:

> *'Designed by British telephone engineer Tommy Flowers, Colossus was built to speed up code-breaking of the complex Lorenz Cypher, used in communications between Hitler and his generals during World War II. Its first job was on 5 February 1944, and its work thereafter is widely thought to have shortened the war and saved countless lives.'*[15]

She also quoted Tim Reynolds, chair of the National Museum of Computing:

> *'Bill Tutte's ingenuity in working out how the Lorenz machine worked without ever having seen it, the skill of those in the Testery who broke the cipher by hand, and Tommy Flowers' design of the world's first electronic computer Colossus to speed up the code-breaking process are feats almost beyond comprehension.'*[16]

By far the greater part of the information concerning the planets, the stars and the galaxies, for more than 2,000 years from antiquity right up to the middle of the 20th century, has come to us in the form of encoded messages of light. By deciphering these messages, astronomers have solved major problems concerning the movements of these bodies, their distances, their temperatures, their chemical compositions and their internal structures, among other properties. The key to breaking these encoded messages was our understanding of the properties of light and the behaviour of matter under a large variety of conditions.

Just as Bletchley Park was the main centre, the laboratory, for deciphering the German codes, so physical laboratories all over the world have provided data that have been the key to unlocking the encoded messages we receive from the cosmos. These data have been combined with mathematical and mechanical models of celestial bodies – and in the 20th century with computer models – to produce what we currently know about the extra-terrestrial universe. In this chapter we will consider the most important properties of light that we have used in this quest.

Light travels in straight lines

The space between the planets, stars and galaxies can be considered, for most purposes, to be very nearly empty space, meaning that light, generally speaking, travels in straight lines. (The bending of light by the Sun, discussed in connection with Einstein's General Theory of Relativity, is a very small effect, so it need not concern us in this chapter. In the final chapter we will see how this bending of light becomes very important, and can be used to make important discoveries concerning the nature of the universe; here we will just accept that, to a very good approximation, light travels in straight lines.)

This fact gives rise to the phenomenon of parallax. In Chapter 1 we saw how Hipparchus used the parallax of our Moon, with respect to the Sun at the time of an eclipse, to measure the distance to the Moon. We also saw how astronomers tried to use the parallax of Venus, with respect to the Sun at the time of a transit of Venus, to measure the distance to Venus. However, the stars are so far away and our Earth is too small to give rise to a measurable shift in their positions if we had to observe the nearby stars, with respect to the more distant stars, as seen from widely different positions on Earth.

For measuring the distances to the stars we can use the method of **stellar parallax**. This method is very similar to that used by surveyors for measuring distances on the surface of the Earth. If a surveyor wants to measure the width of the river she can mark out a baseline on one side of the river and then measure the angles that an object, such as a tree on the opposite bank, makes with each end of the baseline. By drawing to scale the triangle formed by the tree and the baseline she can measure the distance to the tree on her drawing, and then use the scale factor to work out the actual distance to the tree. As we have already noted, the size of the Earth is too small in comparison to stellar distances to use this method. However, the apparent angular movement of the nearby stars against the more distant stars, as observed from opposite points of Earth's orbit, can be measured. The diameter of Earth's orbit (about 186,000,000 miles/300,000,000 km) rather than the size of our planet itself (about 8,000 miles/12,900 km) is the effective baseline in this case.

In many popular science books, the **light year** is used as a basic unit. This is the distance light travels in one year. Since light travels at 186,000 miles per second/300,000 km per second, then in one year it covers a distance of about 6 million million miles (9.6 million million km). Astronomers normally use a unit called the **parsec**; this is the distance to a star for which the angle of parallax in one second of an arc (which is 1/3,600 of a degree). There are 3.26 light years to a parsec. The method of stellar parallax, as used from the surface of Earth, is limited by distortions introduced by our atmosphere to distances of about 150 parsec. This can be improved on by making observations from an orbiting Earth satellite specially designed for the purpose. Such a satellite, called Hipparcos, was launched in 1989. The name is an acronym for 'high precision parallax collecting satellite' but is also a nod to the Greek astronomer Hipparchus of Nicaea. Hipparcos extended the effectiveness of this method out to a distance of about 500 parsec. On 19 December 2013 the European Space Agency launched another satellite, called Gaia, to push this limit out still further into our Milky Way galaxy. It will be a few more years before we have the results of this latest venture.

The inverse square law for light

The inverse square law is a consequence of the fact that light travels in straight lines. In order to explain this effect, imagine we have a light source that radiates light equally in all directions. Now imagine holding a square loop of wire

(10 cm by 10 cm) at right angles to the rays from the source, one metre away from the source. On a white screen, also at right angles to the rays, but two metres away, this hoop will cast a square shadow of 20 cm by 20 cm. This means that the rays passing through the hoop will be spread out over an area four times the area contained within the hoop, so the strength of the light will be weakened. This is the inverse square law. It explains why, if we had two stars of exactly the same intrinsic brightness, but one star was twice as far away as the other, we will detect only a quarter of the light intensity from the further star than we get from the nearer star. Shortly, we will see how this can be used to measure distances to stars.

Colour and temperature

The following is from Thomas Hardy's *Far From the Madding Crowd* (1874):

> *'The North Star was directly in the wind's eye, and since evening the Bear had swung rounded it outwardly to the east, till he was now at a right angle with the meridian. A difference of colour in the stars – oftener read of than seen in England – was really perceptible here. The sovereign brilliancy of Sirius pierced the eye with a steely glitter, the star called Capella was yellow, Aldebaran and Betelgueux shone with a fiery red.'*[17]

The colours of stars are related to their temperatures. The 'steely glitter' to which Hardy refers implies a bluish-white colour, which is consistent with a temperature of about 10,000 Celsius; the yellow of Capella indicates a temperature of about 5,000 to 6,000 Celsius, whereas

Aldebaran and Betelgeuse have temperatures of 4,000 and 3,500 Celsius respectively.

All hot bodies give off radiation at a very wide range of wavelengths, including the colours of the rainbow – red, orange, yellow, green, blue, indigo and violet. However, temperature determines the colour at which they give off more radiation than they do in neighbouring colours. Imagine an old-fashion electric radiator. When it is first switched on in a darkened room, even before it starts to heat up, some warmth will come from it in the form of infra-red radiation. Then, as it heats up, it will first have a dull red colour, followed by a brighter red colour, then orange and when it is as hot as it can get, it will be yellow. The filament of an incandescent light bulb has a whitish colour, because it has a higher temperature still. The law describing this behaviour is called **Wien's Law**.

Stefan's Law

Before discussing Stefan's Law we need to introduce another way of measuring temperature. This is called the Kelvin temperature scale. It is based on the concept of **absolute zero**. Absolute zero is the lowest temperature possible, where nothing could be colder and no heat energy remains in a substance. Absolute zero is the point at which the fundamental particles of nature have no vibration, except at the subatomic level. By international agreement absolute zero is defined as precisely 0 K on the Kelvin scale, or -273.15^{0} on the Celsius scale.

Stefan's Law states that the energy emitted per square metre of a hot body is related to the fourth power of its temperature. In other words, to work out the energy emitted by one square metre of a hot body, we have to take the temperature of the body and multiply it by itself four times.

To illustrate this law in action, consider a body at about 26.85° Celsius and a second at 326.85° Celsius. On the Kelvin scale they will have temperatures of 300 K and 600 K (degrees are not used on this scale) respectively. This means that the temperature of the second is twice that of the first on the Kelvin scale, so the hotter body will give off 16 times more energy than the cooler one, since we have to multiply the number 2 by itself 4 times.

Wien's Law and Stefan's Law, together with the inverse square law, can be used to find the radius of a star. We can measure the amount of light we receive from a star, and if we also know the distance to the star, having measured its parallax, we can work out how much energy is being radiated from its surface in all directions. We use the colour of the star to find out its temperature, and with Wien's Law we can work out how much energy it is giving out per unit of surface area. Since we have already calculated the total energy it is emitting, we can work out its total surface area, and from this calculate its radius. Stars are so far away that even with a very powerful telescope they never can be seen as discs, unlike planets.

What are the stars made from?

French philosopher Auguste Comte (1798–1857) is reputed to have said that because the stars are so far away we would never be able to reach them, so in turn we would never know anything about their chemical composition. Obviously, he was wrong! Although we are still very far from getting to even the nearest star, we do have the means to determine their chemical compositions.

The tool used by astronomers to investigate the chemical compositions of the stars is an important branch of physics known as **spectroscopy**. Initially, this branch started with Sir Isaac Newton. Newton passed a beam of white light through a glass prism, and found that on the other side of the prism, instead of a single spot of white light, he ended up with a stretched-out band of all the colours of the rainbow: red, orange, yellow, green, blue, indigo and violet. William Hyde Wollaston (1766–1828), an English chemist, repeated Newton's work but with some refinements. In an experiment carried out in 1802, he replaced Newton's circular hole with a narrow slot of about 1/20th of an inch wide. The spectrum of light, now free from the overlapping colours, showed him seven lines that were all dark, which he regarded as the natural boundaries between the basic colours.

In 1814 a Bavarian instrument-maker, Joseph von Fraunhofer (1787–1826), invented an instrument consisting of a system of lenses and a prism, which gave a much more detailed spectrum of the Sun. This instrument is now called a **spectroscope**. With this instrument Fraunhofer made the discovery that the rainbow range of colours (called the **spectrum**) had about 600 dark lines crossing it. He noticed that a particular line, which he labelled the D line, in the yellow part of the spectrum seemed to coincide exactly with a bright line that was given off when sodium was put into a flame. These lines resemble, very closely, the lines associated with the bar codes used in most shops to identify different products. As a result of his work, the solar spectrum is now called 'the Fraunhofer Spectrum'.

The subject was taken much further by a Heidelberg chemist Robert Bunsen (1811–99) and his colleague, the physicist Gustav Kirchhoff (1824–87). In the first of their discoveries, which was announced in 1859, they noted that particular sets of lines are associated with individual chemical elements. William Allen Miller (1817–70) of King's College London had already noted the light emitted by electric arcs struck between two metal rods differed depending on the metal used. By making many measurements in the laboratory, Bunsen and Kirchhoff identified particular sets of lines with many different metals, and dramatically confirmed these results by identifying two new elements: caesium and rubidium.

In 1859, Kirchhoff carried out further experiments, which led to an understanding of how the lines were produced. He looked at the spectrum of the Sun through a yellow sodium flame, expecting the bright light of the flame to mask the dark line in the Sun, but instead it became even darker. He deduced, and his interpretation became generally accepted, that sodium vapour existed in the glowing atmosphere surrounding the Sun, and absorbed light of that particular wavelength. Such lines became known as **absorption lines**. We now had a tool for working out the chemical compositions of the Sun and stars.

The application of spectroscopy to astronomy led to the discovery of a new element. In 1868, Pierre-Jules-César Janssen (1824–1907), a French astronomer, discovered an absorption line in the atmosphere of the Sun which he could not link to any known element on Earth. This element was also discovered on the Sun, independently, by the English astronomer Sir Norman Lockyer (1836–1920), who called it helium, after the sun god Helios. Twenty-seven years later the Scottish chemist Sir William Ramsay (1852–1916) discovered this element in a terrestrial laboratory.

Spectra and atomic structure

The link between atomic structure and the spectra of elements was eventually explained by the Danish physicist Niels Bohr (1885–1962), who was the first person to successfully apply a relatively new branch of physics, called **quantum theory**, to a model of atomic structure proposed by Ernest Rutherford (1871–1937).

Although physics has made numerous contributions to our understanding of astronomy, astronomy has also provided us with a few very important discoveries. We have already seen that Isaac Newton's laws of motion and law of gravitation were inspired by his work on the orbits of planets and comets, so here we have two examples of astronomy acting as a spur to physics. We also saw that the fact that light has a finite speed was a discovery made by the astronomer Rømer, but it became a cornerstone of Einstein's special theory of relativity.

In his inaugural lecture as Jacksonian Professor of Natural Philosophy at Cambridge University, Professor

A. H. Cook drew attention to the contribution made by astronomy to specific spectrographic problems. He concluded his lecture with these words.

'Astronomical observations have been crucial in the development of the atomic spectroscopy, and I believe that the reason is that astronomical sources afforded conditions which were not at that time accessible in laboratory sources.'[18]

We can see this in the work of a physicist called John William Nicholson, who published a series of papers arguing that the most primitive forms of matter existed in such objects as **gaseous nebulae** (large gas clouds between the stars) and the **solar corona** (the very hot extended atmosphere of the Sun). He believed that the atoms of special elements in these objects consisted simply of a ring of a few electrons surrounding the positive, but very small, nucleus. He also applied the ideas of early quantum theory to these hypothetical atoms, and was able to account for some of the spectrum lines in the solar corona and in nebulae. This was before Rutherford conceived his own ideas on atomic structure. Bohr's work on the subject built on Rutherford's idea, but his first papers on the subject he acknowledged, corrected and extended the pioneering work of Nicholson.

Before we discuss the Bohr-Rutherford model of the atom, we need to mention the work of Max Planck (1858–1947) on the quantum theory of radiation. Planck was trying to explain Wien's Law and Stefan's Law, but found he could only do so if there was an absolute minimum of energy associated with each colour of light. In other words that there was a smallest 'packet of energy' linked to each and every colour, which was different from one colour to the next. This smallest packet of energy he called a **quantum**.

In 1911, Ernest Rutherford, in the Schuster Laboratory at Manchester University, with two of his students, Hans Geiger and Ernest Marsden, carried out a series of observations on which he was to base his model of the atom. He proposed that the atom consisted of a dense central nucleus, in which the major part of the mass was concentrated, and which had a positive electric charge; and this was surrounded by a number of negatively charged electrons, which were orbiting the central nucleus. He believed that in some ways it was similar to the Solar System, with the planets orbiting the Sun, but whereas gravity was the controlling force in the Solar System, it was electrostatic attraction in the case of the atom. Unfortunately, according to the laws of physics as they were understood at that time, the electrons would radiate energy and they would spiral into the nucleus as a result of this energy loss.

Bohr joined Rutherford in Manchester and the collaboration between the two men led to an improved model of the atom, which became known as the Rutherford-Bohr model. In this model the electrons orbited in orbits at specific distances from the nucleus. Bohr's lanes, or orbits, were circular and the electrons had to obey strict lane discipline. In such a lane an electron would not radiate energy. An electron can absorb radiation and move from a lower orbit (one closer to the nucleus) to a higher one (further from the nucleus). As an electron moves from a higher orbit to lower one, it will radiate energy. Each line of radiation has a wavelength (the distance between successive peaks or troughs of the wave) and a frequency (the number of peaks that will pass through a given point in a second).

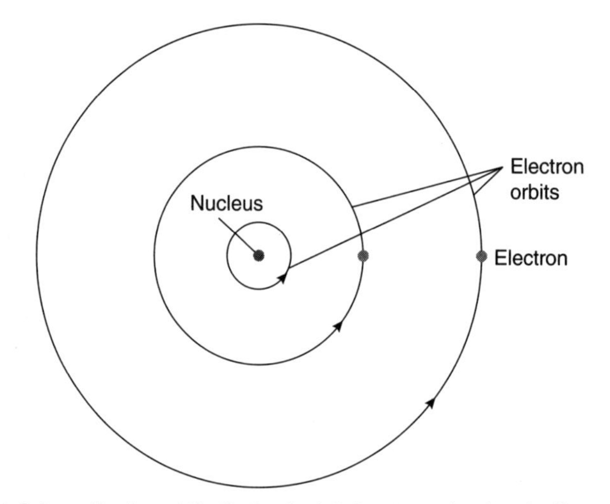

▲ **Schematic view of the Rutherford-Bohr atom, showing the first three allowed orbits.**

In 1913, Bohr wrote three very important papers on atomic structure. In the first, he worked out the spectrum that one would expect from hydrogen. In the second paper, he extended his reasoning to include other chemical elements. And in the third paper he discussed how the chemical elements could be combined to form molecules. When he wrote these papers it was still not clear what the nucleus was made of, but subsequent work showed that two particles existed in the central nucleus. In order to describe how the chemical elements are built up, we will make use of this later understanding of a nucleus consisting of protons and neutrons.

The mass of the proton is about 2,000 times that of the electron, and it has a positive electric charge of equal magnitude (but of opposite sign) to the charge on the electron. The neutron has, very nearly, the same mass as the proton, but it has no electric charge. The hydrogen atom has one proton in its nucleus, and one electron orbiting it. Helium has two protons and two neutrons in its nucleus, and two electrons orbiting the nucleus. As just another example we will describe the element carbon, which has six protons and six neutrons in the nucleus, and six orbiting electrons. It is the orbiting electrons that determine how one atom will combine with another, so they define the chemical properties of an element. However, when an element is heated up it will emit radiation, and the spectrum it will emit is also determined by the orbiting electrons. It is this simple fact that allows us to work out the chemical compositions of our Sun and the stars. The major part of the mass of an atom is determined by the number of protons and neutrons it has in its central nucleus.

The masses of planets and stars

We have already seen that according to Newton's law of gravitation, every particle in the universe attracts every other particle with a force proportional to the product of their individual masses, and inversely proportional to the square of the distance between them. Strictly, the law applies to individual small particles that are point masses. The Sun, stars and the planets are giant spheres, so how does this law apply to them? Newton invented a branch of mathematics, called **integral calculus**, to solve this problem. He was able to show that at any point outside a sphere, the gravitational field, due to all the particles in it, behaved as if all its total mass were concentrated at its centre. We can use this fact to work out the masses of planets and of some special classes of stars.

Finding out the mass of a planet that has a satellite is a simple matter. We have already seen how the scale of the Solar System is determined, so we know the distances to all the planets. This means we can also work out how far a satellite is from its parent planet, and by observing it over a period of time we can work out its orbital speed. In many circumstances, the mass of a satellite is small compared to the mass of the planet, so its movement has only a small effect on the planet itself. The speed of the satellite is due to the gravitational tug of the parent planet, so we can work out the mass of the planet by using its orbital speed and its distance from the planet. The masses of Mercury and Venus, since these planets have no satellites, are worked out from the effect they have on each other's orbits.

Many stars in our Milky Way galaxy are binary stars; that is, they consist of two stars orbiting each other. There are three types of binaries: visual binaries, eclipsing binaries and spectroscopic binaries.

Visual binaries are far enough apart, and close enough to our Earth, to be seen as separate stars. Sometimes one component of such a system is much more massive than the other, so the less massive body will clearly orbit around the other. If the pair is close enough for us to find the distance to them by parallax, we can work out their separation, and we can, over a period of time (normally several years), observe how long it will take for the smaller body to complete one orbit. Even if we are only able to see part of the orbit we can still, sometimes, work out how long it will take to complete the rest. Since the bodies are attracting each other gravitationally we can use the information that we have to work out the mass of the larger body. If the two bodies are comparable in mass then they will move about their common centre of mass. We can use a child's seesaw to illustrate this effect. An adult, sitting near the point of balance can support a small child sitting at the far end, so the centre of mass of adult and child is closer to the adult than it is to the child. Two stars with comparable mass will orbit about their common centre of mass, then it is possible to measure the masses of both components.

An **eclipsing binary** is one in which one star orbits another, but they are too far from Earth to be seen as separate bodies. If the plane of orbit of the smaller member lies close to the line of sight from Earth to the larger, then the two bodies will sometimes eclipse one another. Suppose that at one time both bodies are next to each other, so from Earth a sensitive measuring device will pick up the light from both components. Now consider what happens when the smaller is in front of the larger. At such a time it will blot out some of the light from the larger, and the amount of light we receive will be reduced. At some other time the smaller body will pass behind the larger, so we will only receive the light of the larger member. Thus the light we receive from such a pair will fluctuate and the resulting variation is called the **light curve of the system**. From such a curve we can work out the relative sizes of the stars involved, and we can also, sometimes, make deductions about their masses.

A **spectroscopic binary** is one in which we can make deductions about their relative movements by studying the combined spectra of the pair. If one member of the pair is travelling towards the Earth, then all its spectral lines will appear shifted towards the blue end of the spectrum, and from the amount of shift we can work out its speed towards us. If the other member is travelling away from us, then its spectral lines will be shifted towards the red end of the spectrum.

All three of these methods can be used to deduce the presence of planets around other stars. Planets only reflect light from their parent stars, they have no light of their own, and this is why we generally cannot see them from Earth. However, if a star is seen to move with respect to other distant stars, but has no visible luminous companion, we can surmise that its movement is due to a non-luminous body, such as a large planet. The planets Jupiter, Saturn, Uranus and Neptune actually move our own Sun about the common centre of mass of the Solar System.

The presence of a planet orbiting a distant star can also be deduced if the planet moves between us and the star, thus reducing the light from the star. Alternatively, a star can show periodic changes in its spectral lines, from which we can deduce that sometimes it is moving towards us, and sometimes it is moving away from us. At the moment we know of a few hundred stars that have a planet, or a planetary system, but the search still continues for more evidence for solar-like systems far beyond our own.

Conclusion

In this chapter we have seen how we can measure some of the properties of stars, and how we can work out some of their other properties by decoding the visual messages we receive from them using the known laws of physics as the keys to these codes. In the next chapter we will discuss the most important characteristics of our Sun.

The Sun

A star is drawing on some vast reservoir of energy by means unknown to us. [This reservoir can scarcely be other than the subatomic energy which, it is known exists abundantly in all matter; we sometimes dream that man will one day learn how to release it and to use it for his service.] The store is well-nigh inexhaustible, if only it could be tapped. There is sufficient in the Sun to maintain its output of heat for 15 billion years.[19]

Sir Arthur Stanley Eddington

The external and visible features of the Sun

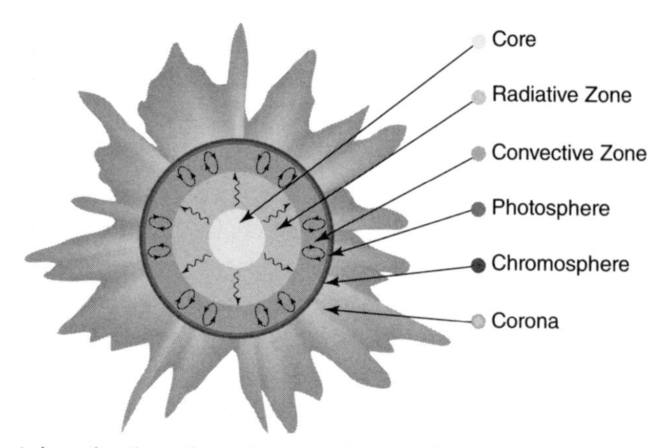

▲ **A section through our Sun, showing its main components.**

The solar diameter is 864,000 miles (about 1,400,000 km), which is 109 times the diameter of Earth.

The light from the visible disc of our Sun is called the **photosphere**. Above this is the **chromosphere**, then the **transition region** before we get the vast extended **solar corona**. The radiation from the chromosphere and the corona is difficult to see against the light of the photosphere, so we can only see these components of the Sun when the photosphere is obscured by the Moon, during a total eclipse of the Sun.

Because the photosphere consists of highly ionized gases (the atoms of which have lost many of their electrons), it is opaque and we cannot see below the photosphere. In other words, although we see the glowing gases of the photosphere itself, we cannot see below its surface. Underneath this is the **convective zone**, which gives rise to the solar granulation that makes its surface look like a bowl of porridge being heated on a cooker. The chromosphere is roughly 1,200 miles (2,000 km) thick. The transition region is thin, very irregular and it separates the chromosphere from the corona. The corona of the Sun is a very tenuous and very hot gaseous region that reaches to the outer edges of the Solar System. In places it has a temperature of about one million Kelvin. Astronomers are not really sure why it is so hot, but one theory is that sound waves, which originate in the chromosphere, develop into shock waves in the rarefied gases of the corona, and it is these shock waves that cause the heating of the corona. It is rather like the waves of the ocean developing into breakers – which are really shock waves in water – as they encounter the beach.

Sunspots and the solar cycle

It is generally believed that sunspots were first seen, with the aid of a telescope, by Galileo in 1610, and ever since they have been of intense interest to astronomers. Although they appear to be very dark regions of the photosphere, this is entirely a contrast effect. The region of the sunspot is cooler than the surrounding areas of the Sun, and it emits less light and as a result looks darker. However, most sunspots are as bright as the full Moon. Observations on the longer-lasting spots show that the Sun is rotating, but not as a solid body. The rotation rate near the poles is about 37 days, but it is about 26 days in the equatorial region.

The number of sunspots also varies and reaches a maximum roughly every 11 years, but the period between successive maxima can be as short as 7 or as long as 17 years. The cycle begins with the formation of spots in middle latitudes of each hemisphere, about 40 to 50° from the solar equator. Subsequently, spots form at lower latitudes until most of the surface is covered. After maximum coverage it is the high latitude spots that disappear first, and the last to fade completely are those closest to the Sun's equator.

The magnetism of the Sun

In 1896, Dutch physicist Pieter Zeeman (1865–1943) made the discovery that the spectral lines given off by different elements were split into several components in the presence of a magnetic field. He also showed that the separation between the split components was related to the strength of the magnetic field. This meant that it was possible to measure the strength of a magnetic field by measuring the separation between the splitting of the spectral lines of different elements. In 1906 the American astronomer George Ellery Hale (1868–1938), working at Mount Wilson Observatory, used this effect, the Zeeman Effect, to measure the magnetic field in sunspots.

Hale was able to show that in sunspot pairs, the magnetic field would emerge from one member of the pair and re-enter the Sun at the other member, so the pair behaved, magnetically, as if there were a bar magnet beneath the Sun's surface. Usually, because of the rotation of the Sun, one member would be slightly ahead of the other, so it was called the leading member. Hale showed that in the Northern hemisphere of the Sun, the leading member of a pair would have a different polarity to that of a pair in the Southern hemisphere. Moreover, the polarities would switch in the build up to solar maximum, from one maximum to the next. Subsequent observations were able to show that the Sun also had magnetic poles, close to its rotation axis, similar to those of Earth, but the fields at the poles were much weaker than the fields in sunspots. They also switched at the start of a new cycle.

It is now generally believed that the combination of the Sun's rotation, and the convective motions in the convective zone, twist the magnetic field lines into braided ropes of very intense magnetic fields. These ropes will rise to the surface to form a sunspot pair, linked together by a looped prominence, in which we have streaming motions of the ionized gases of the Sun.

The interplanetary magnetic field and the solar wind

The loops of the solar magnetic field, which arch into the corona and are anchored in the active regions of low latitudes, will be stretched out near the plane of the Earth orbit into interplanetary space by the outflowing solar wind. As the loops move out radially the Sun rotates about its axis, leading to the winding up of the stretched-out field lines. As a result, the interplanetary field near the plane of the Solar System assumes a spiral configuration. This is called the **garden hose effect**, because the lines of the magnetic field, like the waterjet of a rotating garden hose, form a curved spiral, but the solar wind particles, like the droplets of water, always move in a radial direction away from the Sun. Besides having a spiral form the interplanetary field is divided into four sectors, with the field pointing in opposite directions in each of the sectors.

The radiative zone and the reacting core

In the convective zone heat from the interior is transported out by means of convective currents, in which hot gases rise to the surface, emit radiation, cool down and then return in a constant ongoing process.

Beneath the convective zone is the radiative zone, where heat from the solar interior is transported outward by radiation. A packet, or quantum, of energy from the interior, called a **photon**, does not come out in one straight ray. It follows what is sometimes called a 'drunkard's walk', in which the photon in constantly being absorbed and then re-emitted by atoms. A single packet of energy will take about 170,000 years to travel from the reacting core to the convective zone.

The Reacting Core

The energy of the Sun is generated in the reacting core in which hydrogen is converted into helium by means of nuclear processes. Before going on to discuss how this happens, we need to introduce the concept of isotopes of elements.

Isotopes

An isotope of an element has the same number of protons in the nucleus, and the same number of electrons orbiting the core, as the parent, or main version, of the element, so it has exactly the same chemical properties. However, the daughter isotopes differ from the parent in the number of neutrons in the nucleus, so the atomic mass of the daughter is greater than that of the parent. Hydrogen has three isotopes: hydrogen, deuterium and tritium. Hydrogen has one proton on the nucleus and one orbiting electron; deuterium has one proton and one neutron, and tritium has one proton and two neutrons.

Hydrogen is converted into helium in the reacting core of our Sun by means of a process called the **proton-proton chain**. In the core of the Sun the temperature reaches 15 million Kelvin, so all the hydrogen atoms have been stripped of their electrons. At these high temperatures, the protons of the hydrogen nucleus are moving so fast that they can overcome the electrostatic repulsion that exists between similarly charged particles, for nuclear forces to take over. Two protons will fuse together and give off a positively charged electron in the process, so we have one nucleus of deuterium. Another proton can collide with this deuterium nucleus to form a short-lived nucleus of an isotope of helium, called helium-3. Helium-3 nuclei are not normally found on Earth, but they can exist, for short periods, in the interiors of the Sun and stars. Two helium-3 nuclei can then combine to produce one nucleus of helium-4 (the standard form of helium) plus two protons. The mass of one helium nucleus is less than that of four protons, so in this process some mass has been lost. According to Einstein's special theory of relativity, a small loss of mass releases an enormous quantity of energy. It is this energy that fuels the Sun, the stars and the hydrogen bomb.

Our Sun has been producing energy in this way for 5,000,000,000 years, and will continue to do so for the same amount of time. The realization that nuclear energy is important to astronomy provided the final stages of continuing arguments concerning the ages of the Earth and Sun, which started at the end of the 19th century.

The great debate on the age of the Sun and Earth

When geologists started studying the formation of geological features they came to the conclusion that a very long time span was necessary for geological processes to have formed the features. When Charles Darwin (1809–82) published *On the Origin of Species*

(1859), he made it clear that millions of years would be needed for the evolution of higher forms of life from the more basic forms by the processes of natural selection. Darwin said:

'He who has read Sir Charles Lyell's grand work on The Principles of Geology... *and yet does not admit how incomprehensibly vast have been the past periods of time, may at once close this volume.'*[20]

Darwin tried to get some estimate of the age of the Earth by calculating the denudation of the Weald in south-eastern England. By comparing the volume of material eroded from the dome with an estimate of the rate at which marine denudation would have removed it, he obtained an estimate of 300 million years.

One very influential physicist who did not like Darwin's estimate was William Thomson, 1st Baron Kelvin (1824–1907). He tried to calculate the age of the Sun. From his calculations, he concluded that the Sun had illuminated the Earth for no more than 100 million years, and said:

> *'What then are we to think of such geological estimates as 300,000,000 years for the "denudation of the Weald"? Whether it is more probable that the physical conditions of the Sun's matter differ 1,000 times more than dynamics compel us to suppose they differ from those of matter in our laboratories;*

or that a stormy sea with possibly Channel tides of extreme violence, should encroach on a chalk cliff 1,000 times more rapidly than Mr Darwin's estimate of one inch per century.'[21]

By the end of the 19th century there was a great deal of opposition to Kelvin's ideas from geologists. One, Professor Thomas Chamberlin (1843–1928) from the University of Chicago, went so far as to speculate that there might be other sources of energy that had not yet been discovered, locked up in the interior of the Sun.

We thus see that a geologist was willing to speculate that the long timescales required by observations in geology and biology were actually telling us something about the nature of the physical world, which had not, as yet, been discovered by the physicists themselves. We now know that Chamberlin was right and that these words foreshadowed the discovery of nuclear energy.

The structure and evolution of the stars

The inside of the star is a hurly-burly of atoms, electrons and ether waves. We have to call to aid the most recent discoveries of atomic physics to follow the intricacies of the dance. We started to explore the inside of the star; we soon find ourselves exploring the inside of an atom.[22]

Sir Arthur Stanley Eddington

In this chapter we will consider the structure and evolution of the stars, but we will begin by discussing how stars are classified.

The classification of stars

The absorption lines crossing the continuous spectrum (the rainbow-like range of colours) can be used to classify stars. Originally, the varying strength of a particular set of lines in the hydrogen spectrum, called **Balmer lines**, was used to classify the stars into classes labelled alphabetically from A to P. A stars had the strongest hydrogen lines. Soon after this system had been introduced, it was discovered that the spectral lines of the various stars have widely differing strengths in stars with different temperatures. As a result a new system was introduced, which ordered the stars in order of decreasing temperatures. This led to the rearranging of the alphabetically labelled classes and the dropping of other classes.

Most stars can be classified under seven spectral types, labelled by the letters O, B, A, F, G, K and M. These stars are called the **main sequence**, and with main sequence stars we have a very definite link between their brightness and their temperatures. The order of this classification may be remembered using the mnemonic: 'Oh be a fine girl, kiss me.' The first person to propose this mnemonic was Princeton astronomer Henry Norris Russell (1877–1957), and it still remains popular; a modern alternative is: 'Only boys accepting feminism get kissed meaningfully.'!

The spectral types can be further subdivided into sub-classes, denoted by the numbers zero to nine, placed after the spectral type letter.

Stars also vary a great deal in absolute brightness, or, to use a more objectively measurable quantity, luminosity. The luminosity is defined as the total energy radiated per second from a star. The **Hertzsprung-Russell Diagram** shows the relationship between the spectral type and the luminosity of a star. The brighter stars are also the hottest stars. The luminosity of the star is also related to its mass; more massive stars being more luminous than less massive ones.

The internal structure of a star is also related to its mass. For our Sun and stars of similar or lower mass, about 1 per cent of the mass is in the form of convective envelopes, which occupy the outermost shell of the star. For stars between 15 and 1.5 **solar masses** (a solar mass is equivalent to the mass of our Sun), between 38 per cent and 6 per cent of the mass is in the form of convective regions.

Special types of star

Although most stars spend the major part of their lifetimes on the main sequence, there are other types of star.

Red giants

These stars are much larger than our own Sun, having diameters varying from 10 to 1,000 times that of the Sun. Their surfaces have temperatures that range from 2,000 to 4,000 Kelvin. They are one of the stages near the end of a star's life, after it has moved off the main sequence. The core now consists of inert helium, and the conversion of hydrogen to helium starts in a spherical shell just outside the core.

At later stages in red giants, helium can be converted to still heavier chemical elements by a process called the **carbon-oxygen cycle**. Three nuclei of helium can fuse together to make one atom of carbon. When another helium nucleus collides with a carbon nucleus it will fuse to form oxygen, and with yet another helium nucleus we can get to neon. This type of process can continue until we get to iron, but it is impossible to get energy by fusing iron to form yet more massive elements because these reactions must have additional energy delivered from another source.

White dwarfs

Before discussing the structure of white dwarfs, we need to clarify something about the type of pressure that supports a star against the tendency of any large collection of particles to contract under gravity. There are three main types of pressures involved in supporting the mass of a star.

The first is called **gas pressure** and it comes from the fact that all atoms and molecules of a gas, at any sort of temperature, are moving about at high speeds and these high speeds are what cause the gas to have pressure. This type of pressure plays a major role in supporting most stars against gravitational contraction.

The second type of pressure is called **radiation pressure**, and it arises from the fact that the packets of light energy, the photons, as they move about inside a star also exert pressure. This type of pressure plays a less important role in most stars.

The third type of pressure is called **degeneracy pressure**. Degeneracy is a quantum mechanical effect, and in order to understand the concept it is necessary to explain two fundamental principles of quantum mechanics.

The first is called the **Pauli exclusion principle**. According to this principle, only two electrons, spinning about their own axes in opposite directions, are allowed to exist in a very small volume of a theoretical space, called **phase space**. The size of this space is determined by **Heisenberg's uncertainty principle**. According to this principle it is impossible to know the position and the speed, or more precisely the momentum, of a particle with the same accuracy at a given time. In other words, if we know the position of a particle, then we cannot know much about its momentum. As a consequence of this law, if the electron is confined to small region of space, its position will be known with great accuracy, but little will be known about its momentum; it will be moving about in all directions, at high-speed. As a result of this random motion the electron exerts pressure. This is called **electron degeneracy pressure**. It is this pressure that prevents a white dwarf from collapsing under gravitational forces.

In a white dwarf the electrons have been stripped from their nuclei and all the particles pack together very tightly, so the internal pressure becomes very high from the degeneracy of the electrons. The star will try to contract under the attractive forces due to gravity, but this is balanced by the degeneracy pressure. Theoretical calculations on the structure of white dwarfs show that such stars cannot have a mass exceeding 1.4 solar masses. These stars have a very high density. In particular a white dwarf is half as massive as the Sun, yet only slightly bigger than our Earth. An Earth-sized white dwarf has a density of about 1,000 million kilograms per cubic metre. Earth itself has a density of about 5,000 kilograms per cubic metre. This means a white dwarf is 200,000 times as dense as Earth.

Neutron stars

A neutron star is a very small and extremely dense star made almost completely of neutrons. It is really a very large nucleus held together by the force of gravity. The radius of a neutron star is about 10 km, or roughly 6 miles. It has a mass of about 1.4 to 5 times the mass of our Sun. Neutron stars turn on their own axis at very high speeds, taking from 0.001 second to 30 seconds to spin once.

Pulsars are believed to be neutron stars spinning at such high speeds, in addition emitting radio waves due to electrons spiralling in the strong magnetic field of the star. Normally these spiralling electrons produce beams of light and radio waves. When the beam moves past our Earth, it shows up as a pulse, rather like the light from a lighthouse is seen as a flash by a passing ship. These pulses are the reason why these bodies are called pulsars. They were first discovered by Jocelyn Bell (1943–) working at Cambridge University in 1967.

Novae and supernovae

A nova is the result of a surface explosion of material that has fallen onto the surface of a white dwarf star. Nuclear reactions, in which lighter elements are fused to make heavier ones, are started when the white dwarf collects material from its surroundings. The explosion causes a very sudden, but short-lived brightening of the star, and this is what we see as a nova. But the brightening usually happens over a few days and thereafter the star, over a period of weeks or even months, returns to its original brightness. Very often the original star was very faint as seen from Earth, and this is why people thought that its sudden brightening indicated that a new star had been born, hence it was named a nova or a 'new star'.

Supernovae are very different from novae. Supernovae are very massive stars, at least eight times more massive than our Earth, that undergo very sudden and extremely great increases in their brightness in a very short time and then gradually, over a period of days, weeks or months, return to something like their original brightness. Sir Fred Hoyle (1915–2001) was one of the first astronomers to realize that supernovae played an important part in the chemical evolution of the universe. We'll discuss this further in the next chapter.

Stages of stellar evolution

Stars are formed from the collapse, under gravitation, of vast gas and dust clouds that exist between the stars. When such a cloud, many hundreds or thousands of times the mass of our Sun, contracts it will fragment into smaller clouds with masses that can range from the Sun's mass to masses a few times the mass of the Sun. As these smaller clouds collapse even further, the pressure, temperature, and density of the centre of the cloud will increase considerably until the temperature is high enough for nuclear reactions to occur. At this stage the **proto-star** becomes a main sequence star. During this stage the star is mainly fuelled by converting hydrogen into helium. However, other nuclear reactions occur in some of the heavier stars.

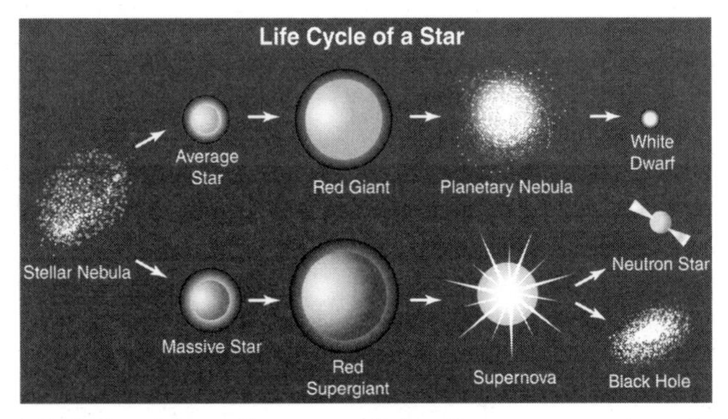

▲ **Evolution of stars – the upper path denoting an average star and the lower path a massive star.**

Once all the hydrogen in the core has been converted into helium, the energy generation required to prevent the star from collapsing under the force of gravitation ceases. At this stage, this core collapses until sufficiently high temperatures are reached to burn hydrogen in a shell surrounding the now inactive helium core. This is called the **shell burning phase**. The outer envelope of the star expands considerably and cools as it does so. As a result, the star will move off the main sequence to become a red giant. Although the outer envelope of the red giant expands considerably, the core itself contracts a little and becomes more dense.

In the red giant phase, hydrogen burning (the conversion of the hydrogen into helium), continues in spherical shells further and further from the centre, and the resulting helium is added to the core. The degenerate matter in the call contracts still further, and heats up until the temperature of 100 million Kelvin is reached. At this point, helium burning begins in which helium is converted to heavier chemical elements, such as carbon. Successive nuclear fuels are used up until iron is created. Making elements heavier than iron from those lighter than iron does not yield further energy, but uses energy. When all the nuclear fuels have been used up, there is no energy to support the star against the force of gravitation. At this stage the star will contract to become a white dwarf. A neutron star will be formed if the mass is greater than 1.4 solar masses.

A collapsing star becomes a black hole when its radius has shrunk below a certain limit known as the **Schwarzschild radius**. A black hole has a very high density, so we have a large mass in a small volume. If a photon of light is emitted from the surface of a black hole it will go up, but then it will drop back to the surface.

This distance defines the radius of the **event horizon**. It is fixed with respect to the centre of the black hole. There is no pressure that can prevent the black hole from continuing to contract, so the surface of the black hole recedes from the event horizon but no information can escape outside of this horizon.

Although the total mass of the star is an important contributing factor that determines the final evolution of a star, the details are by no means clear. Theories that attempt to deal with steady mass loss from stars, catastrophic mass ejection and supernovae explosions are still very much in their infancy.

Our Milky Way galaxy

The galaxy is, in fact, nothing but a congeries of innumerable stars grouped together in clusters. Upon whatever part of it the telescope is directed, a vast crowd of stars is immediately presented to view.[23]

Galileo

The Milky Way galaxy is the most obvious feature of the night sky, especially when there is no Moon visible. It consists of about one hundred thousand million stars; our own Sun and all the constellations that are familiar to many, as well as all the other stars that we see on a clear night, far away from the lights of a town or city, belong to the Milky Way. Since all the components of our galaxy are so much closer to us than those of any other galaxy, they have been studied in more detail. Our knowledge of the Milky Way is so vast that it is practically impossible to do it justice in a short chapter, so we have to be very selective. The main aim of this chapter is to give some understanding of the chemical composition of its various components, and to see what this tells us about the chemical evolution of the universe as a whole.

The components of the galaxy

The most obvious feature of the Milky Way is that it consists of a very large number and great variety of different types of stars. Galileo was the first astronomer to use the telescope for astronomical purposes and with this he was able to show that the Milky Way consisted of a vast number of stars. The next astronomer to attempt to determine its structure was William Herschel, who counted the number of stars he could see in different directions. By assuming that all the stars were as bright as the Sun, and that their differences in brightness as

▲ A schematic view of our Milky Way galaxy. as seen from above.

seen from Earth were solely due to the fact that they were at different distances from us, he worked out a model for the shape of our galaxy. He came to the conclusion that the stars were distributed in the form of a flattened disc, with our Sun very near the centre. What prevented Herschel from seeing the full extent of the system was another, invisible, component that cut down the light from the more distant stars and blotted out the light from the most distant stars altogether. This component was very small dust particles, which although small and at a very low density per cubic metre, nevertheless, over the vast distances between the stars, built up to form an **interstellar fog** that cut down visibility near the plane of the Milky Way. When large telescopes were used on the spaces, near and between the stars, they revealed that there were clouds of glowing gases present in what became known as the **interstellar medium**. We now had three components to the Milky Way: stars, dust and gas.

Radio astronomy was to make a big contribution to our study of the spaces between the stars. In 1933 Karl Jansky (1905–50) announced the discovery of radio waves coming from the galaxy. This work resulted from his search for interference on short wavelength radio communications, which he had undertaken at the Bell Telephone Laboratories in the US. Grote Reber (1911–2002), an amateur radio enthusiast, was inspired by the work of Jansky and built a radio telescope in 1937. With this instrument he made the first map of the sky at radio wavelengths. From his study he deduced that radio waves do not follow the same patterns as the stars, so they must be coming from the spaces between the stars.

We now know that although part of the radio radiation is coming from gas between the stars, some of it is coming from very fast moving electrons, which generate radio waves as they spiral around the magnetic lines of force that thread their way between the stars. These fast-moving electrons are some of the remnants produced by the giant supernovae explosions mentioned in the last chapter. Some of the other remnants, which are protons and the nuclei of atoms (also moving at high speed) find their way to Earth, and we call them **cosmic ray particles** – another component of our galaxy.

In 1944 Dutch astronomer Hendrik van de Hulst (1918–2000) made the prediction that in the very low gas densities of neutral atomic hydrogen in interstellar space, hydrogen would emit a radio wave of 21 cm long (about 8 in). This line arises in the following way. The proton and the electron behave as if they are very tiny magnets, with north and south poles. In the hydrogen atom, as the electron orbits the nucleus, the south pole of the electron can be opposite to that of the south pole of the single proton in the nucleus. Since 'like' poles tend to repel, this is an unstable state, so the electron will tend to flip over, and as it does so it emits the 21-cm hydrogen line. We are all familiar with the fact that radio waves are not affected by foggy conditions. Because radio waves are longer than light waves, they are able to penetrate fog, so the radio waves from the galaxy are not affected by the dust particles of the interstellar fog. This means that radio astronomers can see deeper into the Milky Way than optical astronomers can. The 21-cm hydrogen line has been used to map the distribution of neutral atomic hydrogen in our stellar system. On the surface of Earth and in terrestrial laboratories we normally encounter molecular hydrogen, which consists of two atoms of hydrogen; atomic hydrogen, can, however, be found in the low densities of interstellar space.

Stellar populations

In order to discuss the chemical evolution of the galaxy, it is necessary to introduce the concept of stellar populations. The members of a stellar population differ in a few respects, including, physical and chemical characteristics, the way they move and the parts of the galaxy that they occupy. The concept can be clarified by using an analogy. In a large city we have subsets of people, the members of each subset, which can be called a population, will differ in a few respects. For example, they may have different racial origins, religious and cultural affiliations, they behave differently, eat different foods, drive different cars, live in different types of houses in different parts of the city.

There are two main populations in the galaxy, Population I and Population II.

Population I

These objects are distributed in the form of a very flattened disc with a bulge towards the centre, rather like a thin pizza. An important concept here is metallicity, which is defined as the proportion of matter in an object which is in the form of chemical elements that are heavier than hydrogen and helium.

These population I objects are among the youngest in our stellar system. Within this disc, the youngest stars form a subset called the **spiral arm population**, which has comparatively high metallicity. All the population I disc stars move around the centre, sticking fairly close to the central plane, in orbits that are slightly elliptical.

Population II

These objects occupy a very large ellipsoidal volume far above and below the plane of the galaxy. They have highly elliptical orbits, which carry them far from the plane, about the galactic centre. They are much older than the Population I stars, have low metallicity and many of them form tight groups called **globular clusters**. It is believed that they were the first stars to be formed in the early stages of the Milky Way.

The size of the galaxy

Cepheid variable stars form a class that is very useful for determining the size of the galaxy and for finding the distances to other galaxies. The light emitted by these stars varies in a periodic way – the period of the fluctuations varying from a few to many days.

The American astronomer Henrietta Swan Leavitt (1868–1921) looked at many Cepheid variable stars in the **Small Magellanic Cloud**, which is a satellite galaxy to our own. Because these stars are a long way from Earth, but their distribution in space is relatively small compared to their distance from us, she found a relationship between their periods and their overall brightness. The first Cepheid variable detected, Delta Cephei, was close enough to have had its parallax measured, so this provided astronomers with a method of measuring the distance to other Cepheids. The rate of pulsation will give us a measure of its actual brightness, and by comparing this with its brightness measured on Earth we can work out its distance from us, using the inverse square law mentioned earlier. Initially, this method was not very accurate, but it has been refined over the years.

Globular clusters of stars in the **halo** – a roughly spherical volume surrounding the whole of the Milky Way – were used by the American astronomer Harlow Shapley (1885–1972) to estimate the size of the galaxy. He used three different methods to do this. First, some of the clusters contain Cepheid variables, so he could use the method suitable for these stars to measure their distances. Second, he made several assumptions, one of which was that the brightest stars in each cluster all have the same actual brightness, and so by comparing the brightest stars in a cluster of known distance with those in one of unknown distance, he was able to say how far away the latter cluster was. Third, he assumed that all the clusters were of the same actual diameter, and that the origin of the differences in angular size, as seen from the Earth, arose purely from the differences in their distances from us.

The last two methods may seem crude, but the three methods used in combination gave Shapley a very good estimate of the size of the galaxy. His methods worked because he was using clusters of stars out of the plane of the galaxy. (We have already seen that the small dust particles within the plane act like a fog, cutting down visibility in the plane, making it difficult to measure distances accurately. However, the interstellar fog had little effect when looking out of the plane.)

So what is the size of our galaxy? We have already seen that a parsec is the distance measure used by astronomers, which is about 3.26 light years. We now introduce the term **kiloparsec**, which is 1,000 parsec. The diameter of the Milky Way is about 35 kiloparsec, and the distance of the Sun from the centre is about 8 kiloparsec. This means that our Sun is about halfway out from the centre, towards the outer edge of the system.

The great debate on the nebulae

In historical terms the great debate on the nature of the nebulae was a turning point in astronomy. The debate took place on 26 April 1920 in the Baird Auditorium of the Smithsonian Museum of Natural History. It was a debate between Harlow Shapley and Heber Curtis (1872–1942), and it was in essence an argument about the size of the universe. The spiral nebulae played a major part in this debate; these are amorphous and hazy whirlpool-like patches of light seen by many astronomers.

Using the methods discussed in the last section, Shapley believed that the galaxy was the whole of the universe. He not only believed that the galaxy was very large – 300,000 light years in diameter – but he also showed the Sun was not at the centre. Furthermore, he argued that the spiral nebulae were nearby clouds of gas located well within the Milky Way.

Curtis did not accept the size of the Milky Way obtained by Shapley, and showed evidence that the optical spectrum of the spiral nebulae were indistinguishable from the spectrum of our galaxy. They had a similar appearance, with a central belt of obscuring material just as in our own galaxy. Curtis argued that the spiral nebulae were large collections of stars comparable in size to our Milky Way, but located far beyond the boundaries of our own galaxy.

As it turned out, the size of the galaxy obtained by Shapley is far closer to our modern measurements than that estimated by Curtis, but Curtis was right about the spiral nebulae being galaxies in their own right similar to our own.

The mass distribution of the galaxy

Our galaxy does not rotate like a solid body. An ordinary wheel rotates like a solid body, and in this case the speed with which different parts of the wheel rotate are directly related to their distance from the centre of rotation, the axle. This means that parts of the wheel close to the axle rotate more slowly than those on the circumference. Our galaxy does not rotate like this.

The planets of our Solar System obey Kepler's third law of planetary motion, because the major part of the mass of the Solar System is concentrated in the Sun. The planets have some small effect on each other's orbits. With our galaxy the matter is very different, because all the hundred thousand million stars within the galaxy contribute to the total gravitational field. This means that every star really moves under the gravitational attraction of all the other stars in the galaxy. Using the Doppler Effect and the 21-cm hydrogen line astronomers have been able to work out the speeds with which different parts of our galaxy rotate around the centre.

This is called the **rotation curve of the galaxy**. Since there are about one hundred thousand million stars in our galaxy, it would be impossible to work out their individual orbits, even with the largest computer in the world. In order to work out how mass is distributed in the galaxy, astronomers resort to constructing mathematical models of how the 'smeared out mass' could be distributed. They then work out how well the rotation curve predicted by these models fits the observed rotation curve. In doing just that a problem arose that has not yet been fully resolved.

In the outer edges of our galaxy, where we can see hardly any stars, the few stars that are there are rotating as if there was still some matter beyond them. It is rather like the story of the discovery of Neptune revisited – except there are no visible Neptunes beyond the outermost stars. This was the first indication that there is dark matter beyond the outer edges of what we can see. We'll return to this topic in the next chapter.

The link between the dynamical and the chemical evolution of the galaxy

A generally accepted theory of how our galaxy evolved goes like this. All the matter in the galaxy started off as an enormous and very massive cloud that had some rotation. It started to contract under the force of gravity, and as it contracted it began to spin faster. This is due to the **conservation of angular momentum**. We can illustrate this concept by considering an ice skater, spinning on her skates with her arms outstretched. If she now brings her arms in towards her body, the spin rate increases – because the mass distribution has contracted the spin rate must increase.

The spinning slows down the contraction at right angles to the axis of a galaxy, but has little effect on the matter parallel to the axis. This means that the spherical cloud will assume an ellipsoidal shape. As the contraction continues the very big cloud will fragment into smaller clouds, which will then continue to contract. These smaller clouds will become, eventually, the globular clusters of stars, but only after further contraction has occurred to produce the proto-stars within each cluster. The very large stars in each cluster, those of more than eight solar mass units, will evolve very quickly, synthesizing all the elements up to and including iron before erupting in a supernova explosion, the sheer violence of which will produce the heavier chemical elements. These elements will enrich the gases left over, after the formation of the globular clusters, so the rest of the gas will continue to fall towards what will become the disc of the galaxy. This means that during the next stage of stellar formation, the newer stars will be born out of material richer in the heavier chemical elements. Because the material in the plane is rotating at different speeds at different distances from the centre, a spiral density wave is set up, and during the next phase of star formation there will be a still richer interstellar gas, due to second generation stars exploding as supernovae, so the third generation will be born out of this richer mixture of chemical elements.

The elements that eventually formed the planets, and everything on our Earth, including life itself, were synthesized in this way. This means that we are all made of the stuff of stars!

Galaxies and the universe

The evolution of the world can be compared to a display of fireworks that has just ended; some few red wisps, ashes and smoke. Standing on a cooled cinder, we see the slow fading of the suns, and we try to recall the vanished brilliance of the origin of the worlds.[24]

Georges Lemaître

In this final chapter we will discuss galaxies and the universe. There are several different types of galaxy, and there are many thousands of galaxies in the universe, so, naturally it is a vast subject. This means we have to be very selective in what we cover. The main purpose of this chapter is to discuss what galaxies can tell us about the structure of the universe as a whole and how it began.

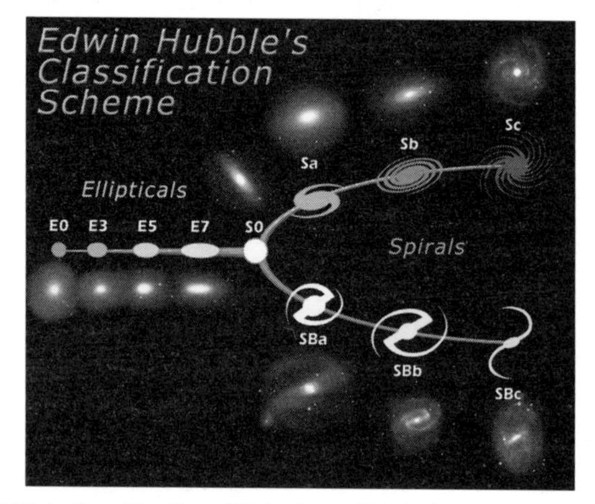

▲ **Hubble's Classification of Galaxies – ellipticals, normal spirals (upper fork) and barred spirals (lower fork).**

Types of galaxy

The system used by astronomers to classify galaxies started with Edwin Hubble (1889–1953). **Normal spiral galaxies** are similar to our own Milky Way, in that they have, roughly, two spiral arms surrounding a compact central nucleus. How tightly the spiral arms are wound varies from one galaxy to another, and leads to further subdivisions within the class. **Barred spiral galaxies** are different from normal spirals in that they have a bar across the centre rather than a nucleus. They also have two spiral arms, but these are not as tightly wound as in normal spirals. The rest of the galaxies are a mixture of other types. **Elliptical galaxies** are, as the name implies, elliptically shaped. They vary from being almost spherical to being highly elliptical, and there are other differences within the class. The **regular galaxies** are rather shapeless. The best-known examples of this type are the large and small Magellanic clouds, which can easily be seen in the night sky in the southern hemisphere. Galaxies are not evenly distributed over the volume of space surrounding us; they are grouped into clusters ranging in size from about twenty to many hundreds. There are also larger groups called super-clusters.

The expansion of the universe

Most people have heard of the 'Big Bang' and many know that all the galaxies are moving away from us at high speeds. Many people have also heard of Hubble's Law, which shows that the speed of recession of a galaxy is directly proportional to its distance from us. Is it right to attribute this law to Edwin Hubble? He was the first astronomer to provide observational evidence for this law, but the concept of an expanding universe did not start with him.

The first person to successfully use Einstein's general theory of relativity to theoretically predict the expansion of the universe was Georges Lemaître (1894–1966).

Lemaître was a Belgian Roman Catholic priest, an astronomer and professor of physics at the Catholic University of Louvain. In 1925 he began to work on a paper that would bring him international fame. This paper, 'A Homogeneous Universe of Constant Mass and Growing Radius Accounting for the Radial Velocity of Extragalactic Nebulae', was published in 1927. In it he deduced the relationship that the speed of recession of a galaxy was directly related to its distance from us. He also, sometime later, introduced the idea that all the matter in the universe was originally in the form of a primeval atom, which exploded as a 'Big Bang', ejecting material into space, but more accurately creating space itself. Many years later he gave the graphic and prosaic description of the origin of the universe (see above).

The 'vanished brilliance'

This 'vanished brilliance' referred to in Lemaître's quote can be seen as what we call **cosmic microwave background** (CMB) radiation. It is the thermal radiation assumed to be left over from the Big Bang. It is fundamental to observational cosmology because it is the oldest light in the universe, dating from the time when particles began to recombine into atoms. With traditional optical telescopes, the space between the stars in the galaxy is completely dark. However, a sufficiently sensitive radio telescope shows a faint background glow, almost exactly the same in all directions, that is not associated with any star, galaxy or any other object. This glow is strongest in the microwave region of the radio spectrum. The CMB was discovered quite accidentally in 1964 by American radio astronomers Arno Penzias (1933–) and Robert Wilson (1936–) and for this they won the 1978 Nobel Prize for Physics. In the early years of the universe the CMB had a temperature of 3,000 K, but because of the expansion this has now dropped to 3 K.

Early in 2013 the European Space Agency reported on the recent results of its Planck space observatory – designed to observe fluctuations of the CMB at microwave and infrared frequencies with very high sensitivity. The report said:

> *'The CMB is a snapshot of the oldest light in the universe, imprinted on the sky when the universe was just 380,000 years old. It shows tiny temperature fluctuations that correspond to regions of slightly different densities, representing the seeds for all future structure: the stars and galaxies of today.'*[25]

The cosmic background radiation and cosmic inflation

The mapping of cosmic background radiation by various satellites gave rise to a problem for the standard model of cosmology. The maps showed an extreme uniformity for this background radiation between the galaxies and the clusters of galaxies. If the universe had started with a 'Big Bang', this uniformity should not have been there. There should have been temperature fluctuations. These were not apparent. Since the universe expanded very rapidly, temperature uniformity could only have been achieved by the transfer of heat by radiation from one part of the universe to the other. Unfortunately, this transfer was limited to the speed of light and cosmologists could not explain how it could have happened. The possible explanation for this came from the work of Alan Guth (1947–), a physicist and cosmologist. In 1980 he proposed that the universe underwent a very violent and extremely sudden period of expansion, which he called **cosmic inflation**. This expansion was actually faster than the speed of light, but it didn't violate Einstein's law that nothing can travel faster than light, because it was space and time itself that was expanding. This scenario was allowed for in Einstein's general theory of relativity.

Another consequence of Einstein's general theory of relativity was that gravitational waves could be generated by parts of the universe. If the theory of cosmic inflation is correct, this rapid expansion would have generated gravitational waves. These waves are like ripples in space and time, and they would have affected the light and other forms of radiation generated at this time. In particular, the gravitational ripples would have polarized the light from the early universe. Radiation from most of the sources, such as light bulbs, candles, car headlamps, and so on, is normally unpolarized, that is, the waves of light are vibrating in all sorts of directions in a random way. If light is passed through a polaroid filter, such as those found in Polaroid sunglasses, such a filter would restrict the plane in which the light vibrates. The light is then polarized. We need to explain a bit more about how gravitational waves give rise to the polarization of light.

If a vertically polarized ray of light encounters an electron, it will cause the electron to move up and down. Gravitational waves behave rather differently. Consider a ring of particles in space. If a gravitational wave encounters such a ring it will distort the ring into an elliptical shape, first with the major axis in a vertical direction. It will then cause the particles to resume their original ring shape, before distorting the ring again in an ellipse with the major axis in a horizontal direction. As the waves pass the particle will oscillate between these three shapes. If there were no gravitational waves from the early universe, electrons will vibrate in random directions, giving rise to unpolarized light. The presence of a gravitational wave will distort the region of space around an electron, causing it to move in a preferred direction. When photons of light interact with such electrons, this preferred direction will be imprinted on the light, giving rise to polarized light.

A team of American scientists set out to measure the polarization of the microwave radiation from the early universe. They set up a special telescope at Earth's South Pole called BICEP2. The name stands for Background Imaging of Cosmic Extragalactic Polarization. They set up their experiment at the South Pole for two reasons. First, because of the clear and dry air found over the Pole; and second, this position also offered an almost direct line of sight out of our own galaxy into deep space. On 17 March 2014 an announcement was made that this polarization had been detected.

Subsequently, astronomers have begun to doubt that this polarized component of the cosmic microwave background is due to gravitational waves. Alternatively, the Milky Way has a large scale, very weak, magnetic field – with large loops of magnetic field lines – which can produce polarization of electromagnetic radiation. Very energetic electrons spiralling around the field lines produce radiation by the synchrotron effect, which is polarization at radio and microwave wavelengths. The magnetic field also causes the specific alignment of non-spherical dust particles. These particles can also produce polarized radiation. Much, if not all, of the polarized radiation could have been produced in this way. Further observations, and different ways of analysing the data, are ultimately needed.

Hubble's Law, radio galaxies and quasars

Early in the history of radio astronomy, small intense sources of radio radiation were found at fixed points. By about 1950 the positions of some of these sources were known with sufficient accuracy for optical astronomers to find them with giant telescopes. Some of the sources were identified with objects in our Milky Way, while others were identified with irregular galaxies that had unusual properties. Radio astronomers soon discovered that the radiation from most of the sources consisted of two or more areas of very strong radio emissions. Radio astronomers had no way of finding the distances to these sources. However, somewhere near the middle of all these objects, were optical sources. The speeds of these sources were measured by the Doppler Effect, and the results showed that they were receding from us at very high speeds. It was then possible to find their distances using Hubble's Law. Since we could find their speeds by the Doppler Effect, we could deduce their distances. They turned out to be a long way from us. The distances to **quasars** were found in a similar way. Quasars are massive and extremely remote celestial objects, which are emitting exceptionally large amounts of energy, but they have star-like images in a telescope. It has been suggested these quasars contain massive black holes and that they represent stages in the evolution of some galaxies.

The age of the universe

Globular clusters can be used to find the age of our galaxy. In the chapter on the Sun we saw that our Sun has been in existence for 5 thousand million years and could live, in its present stage, for a similar length of time. The lifetime of a star depends on its mass. By studying the less massive hydrogen-burning stars in globular clusters, astronomers have worked out the ages of these clusters. These observations give ages of between 11 and 18 thousand million years.

Another way is to use Hubble's Law to work back in time to when all the matter was in one spot. Some cosmologists have come up with an age of 9 thousand million years, which means that the oldest stars in the galaxy are older than the universe itself!

In 2013 more progress was made. Using observations from the Planck Orbiting Observatory, and more refined calculations, the age of the universe is now given at about 13.8 thousand million years. The age of the oldest Population II star, the so-called Methuselah Star, has also been revised on the basis of improved observations and new calculations, so this is now also, roughly, about 14.5 thousand million years old.

Synthesis of elements in the 'Big Bang'

According to many cosmologists the 'Big Bang' was responsible for the formation of most of the helium in the universe, along with small amounts of deuterium and a very small amount of lithium. In addition to these stable nuclei, two unstable radioactive isotopes were also produced – tritium and beryllium. Yet in the last chapter we saw that Population II stars contain traces of metals, so where do they come from? To explain this, astronomers have invoked a pre-galactic generation of super-massive stars, which produced some of the heavier elements, but then were completely ripped apart, so now there is no trace of them. It's another unresolved problem.

Evidence for dark matter

Dark matter is an on-going topic so we will just mention two sources of evidence.

The first is the rotation curve of galaxies. This topic has already been mentioned in connection with our Milky Way, but the rotation curves of a large number of galaxies seem to indicate that the outermost stars are behaving as if there is still more matter, beyond the edges, which has no other consequence except to add to the total gravitational field of these galaxies. This is called dark matter.

The second piece of evidence comes from gravitational lensing. In Chapter 2 we saw how the gravitational field of the Sun can bend light around the Sun. If a distant galaxy is viewed through a nearby cluster, the light from the distant galaxy can be bent by the gravitational field of the cluster itself. Astronomers have found evidence for this which indicates that there is more matter in the cluster than the total masses of all the cluster members. This too is seen as evidence for dark matter.

Evidence for dark energy

The most convincing evidence for an acceleration in the expansion of the universe has come from a type of exploding star known as type 1a supernovae. These are carbon- and oxygen-rich white dwarfs that undergo nuclear explosion. These stars are 40 per cent more massive than the Sun, but they have a radius 100 times smaller. In a binary system the strong gravitational field of the white dwarf can pull off matter from the companion until it exceeds 1.4 solar mass, the limit of a white dwarf. The excess mass destabilizes the star, which then explodes as a supernova. The luminosities of these supernovae are all very similar, so one can use these stars as **standard candles**. This means that by comparing their average luminosity with the luminosity that we can measure on the surface of the Earth, we can work out the distances to these objects.

The results obtained from a few hundred supernovae were astonishing. Astronomers discovered that the fastest-moving supernovae were fainter, and hence further away, than they should have been if the expansion rate was constant or slowing down. In other words, the evidence was consistent with an expansion rate for the universe that was actually increasing. The candidate that is causing this increase is called dark energy.

The most recent estimates for the proportions of the total energy and matter content of the universe in the form of ordinary matter, dark matter and dark energy are as follows: ordinary matter 5 per cent, dark matter 27 per cent and dark energy 68 per cent.

We end with a quote from Edwin Hubble:

> *'From our home on the Earth we look into the dim distances, and we strive to understand the world into which we [were] are born. Our near neighbours we know rather intimately, but with increasing distance knowledge fades, and fades fast, until at the last dim horizon we search among ghostly errors of observation for landmarks that are scarcely more substantial. The urge is older than history. It has not been satisfied and it will not be suppressed.'*[26]

This 100 ideas section suggests ways you can explore the subject in more depth. It's much more than just the usual reading list.

100 IDEAS

5 Constellations Worth Observing

1 **The Plough**. This group of seven stars is part of the constellation of The Great Bear. To many it looks like a saucepan. Its midnight position at different times of the year is as follows: Autumn – low in the North; Winter – East of North; Spring – high in the North; Summer – West of North. The two brightest stars, Dubhe and Merak, point to Polaris, otherwise known as the Pole Star.

2 **Orion the Hunter**. The three stars that form the belt of Orion are easily identified because they are located in an almost straight line, and are very nearly the same brightness as each other. They can be seen in the South during the winter months.

3 **Sirius – The Dog Star**. A line drawn downwards and towards the East, as you face Orion, will take you to Sirius,

which is the head of Canis Major – The Big Dog. Sirius was important to the ancient Egyptians. When it rose in the East just before sunrise, the Egyptians knew that the Nile was about to flood.

4 **Taurus the Bull**. A line drawn through the belt of Orion upwards and towards the West will take you to Taurus – The Bull. Aldebaran, one eye of the Bull, is a reddish star.

5 **The Summer Triangle**. This very big triangle, visible throughout the summer nights, is not a constellation, but its stars are very bright, so they are easily identifiable. The stars that make up the triangle are Deneb, in Cygnus – The Swan, Vega in Lyra, and Altair in Aquila – the Eagle.

5 Objects Seen With Binoculars

If you visit astronomy.co.uk, you will find the sky map option, which will give you the map of the night sky for a particular date. It will show you the constellations that are visible. If you are using a tablet, hold it above your head and it will give you a good idea of the night sky.

6 **The Pleiades** is a group of stars in Taurus. Sometimes they are known as the Seven Sisters. With ordinary binoculars you can see about a dozen stars in the cluster.

7 **The Orion Nebula** is a hazy patch of light just below the belt of Orion. It is a large gas and dust cloud in which stars are being born.

8 **The Crab Nebula** is in Taurus, near the horn of the Bull. It is what is left over after a supernova explosion, so it is called a supernova remnant. This explosion was seen by Chinese astronomers in AD 1054.

9 **The Andromeda galaxy**, is a spiral galaxy that can be seen with the naked eye, but it is better seen through binoculars. Use your downloaded map to find the W of Cassiopeia and

the Great Square Pegasus. The Andromeda constellation is a rather faint one between Pegasus and Cassiopeia. This is where you will find the Andromeda galaxy.

10 **The moons of Jupiter**. The map will show you where to find Jupiter in the sky, and with your binoculars you should be able to find four moons, although, on occasion, one or other of the moons might be obscured by the planet itself.

10 Important Astronomers

11 **Hipparchus** (c. 190 BCE–c. 120 BCE). Ancient Greek who produced the first star catalogue known to the Western World.

12 **Ptolemy** (c. AD 90–168). Ancient Greek astronomer, he promoted a geocentric theory of the cosmos that prevailed for 1,400 years.

13 **Galileo Galilei** (1564–1642). One of the first astronomers to use the telescope for astronomy and with it he made several discoveries.

14 **Sir Isaac Newton** (1642–1727). Among his many achievements he formulated the laws of motion and gravitation and invented the reflecting telescope.

15 **John Flamsteed** (1646–1719). The first Astronomer Royal, based at the Royal Observatory, Greenwich. He was the first to produce a star catalogue based on telescopic observations.

16 **Caroline Herschel** (1750–1848). The first woman astronomer of note. She discovered a few comets and helped her more famous brother William in his work.

17 **Ole Rømer** (1644–1710). Danish astronomer, the first to discover that light travels with a finite speed.

18 **Henrietta Swan Leavitt** (1868–1921). US astronomer who discovered the period luminosity relationship for Cepheid variable stars.

19 **Jocelyn Bell Burnell** (1943–). In 1967, while still a postgraduate, she discovered the first pulsar.

20 **Edwin Hubble** (1889–1953). US astronomer, after whom the Hubble Space Telescope is named, who made numerous contributions to the study and classification of galaxies.

5 DVDs with Astronomical Content

21 ***Carl Sagan's Cosmos*** (1980). Very wide-ranging, but some of the episodes cover astronomy.

22 ***The Ascent of Man*** (1973). Jacob Bronowski. Covers a great deal of science, while a number of episodes cover astronomical topics and aspects of physics relevant to astronomy.

23 ***Wonders of the Solar System*** (2010). Brian Cox. First broadcast on BBC TV.

24 ***Wonders of the Universe*** (2011). Brian Cox. Packed with wonderful visions of the universe.

25 ***Einstein and Eddington*** (2008). Classy BBC dramatization of the testing of Einstein's General Theory of Relativity at the 1919 total eclipse of the sun.

5 Space Movies

26 ***War of the Worlds*** (2005). Steven Spielberg-directed feature that sees Earth invaded by alien fighting machines. Based on the H. G. Wells novel, this film stars Tom Cruise.

27 *Close Encounters of the Third Kind* (1977). Another Spielberg classic where peaceful aliens visit the Earth. Stars Richard Dreyfuss.

28 **2001: A Space Odyssey** (1968). Directed by Stanley Kubrick and based on a story, *The Sentinel*, by Arthur C. Clarke, this intelligent film combines space travel and evolution.

29 *Apollo 13* (1995). Superior drama directed by Ron Howard that is based on the ill-fated Apollo 13 mission to the Moon. Stars Tom Hanks, Bill Paxton and Kevin Bacon as the resourceful crew.

30 *Gravity* (2013). Multi-Oscar-winning British and American 3-D thriller. Stars Sandra Bullock and George Clooney as astronauts set adrift in space.

10 Good Reads

31 *The History of Astronomy: A Very Short Introduction* (2003) by Michael Hoskin. An excellent introduction from a very good professional historian.

32 *The Legacy of Greenwich Observatory* (2012) by Percy Seymour.

33 *A Man on the Moon: The Voyages of the Apollo Astronauts* (1995) by Andrew Chaikin. An epic story very well told.

34 *Galileo's Daughter: A Drama of Science, Faith and Love* (2009) by Dava Sobel. The widely acclaimed author of *Longitude* sets the achievements of Galileo in the context of his time.

35 *Wonders of the Solar System* (2010) by Brian Cox. Well written and lavishly illustrated; based on the BBC TV series.

36 *Astrophysics is Easy!* (2007) by Mike Inglis. A simplified, yet thorough introduction to the essential physics needed to understand the stars.

37 *Rutherford, Bohr, Manchester and the Atomic Nucleus 1911–2011* (2010) by Percy Seymour.

38 *The Milky Way: an insider's guide* (2013) by William H. Waller. A really excellent book about our galaxy.

39 *The Wonders of the Universe* (2011) by Brian Cox and Andrew Cohen. Another excellent book based on the BBC TV series.

40 *A Dictionary of Astronomy* (2012) by Ian Ridpath. Astronomy is full of new words as well as standard words used in a celestial setting. Ridpath's book will help you through the maze.

5 Magazines on Astronomy

41 *Sky & Telescope.* Based in the USA. A good and comprehensive coverage of astronomy and space time.

42 *Astronomy Now.* Based in England and founded by Patrick Moore. Another excellent magazine.

43 *BBC Sky at Night*. Good and reliable information about the night sky. The publication also has a yearbook covering heavens for each month of the year.

44 *Solar Observer*. Published by a group of solar observer enthusiasts, it is the first international magazine on solar astronomy.

45 *Scientific American*. Although it is very wide-ranging, all the articles are well written, and frequently there are articles on astronomy and space.

5 Astronomical Societies to Join

46 **Society for Popular Astronomy**. A national astronomical society, based in the UK, for beginners to amateur astronomy. www.popastro.com

47 **Index of British Astronomical Societies**. Very comprehensive guide to societies all over the country. www.xyroth-enterprises.co.uk/ukas.htm

48 **British Astronomical Association**. This organization was formed in 1890 and has an international reputation for the quality of its observational and scientific work. Membership is open to all interested in taking up astronomy as a serious hobby. www.britastro.org

49 **The Association for Astronomy Education**. It promotes public education in astronomy and supports the teaching of astronomy at all levels of the educational system. The association also runs an Ask an Astronomer question line. www.aae.org.uk

50 **Royal Astronomical Society**. Based in London, this is the senior astronomical society in the UK. Although, in principle, it is open to everyone who can get two fellows to sponsor them, there is also a Friends status. The society has an excellent library in London. www.ras.org.uk

10 Astronomical Sites of Interest on the Web

51 **Hipparchus's measurements of the distance to the Moon**. www.phy6.org/stargaze/Shipparc.htm

52 **A Greenwich guide to measuring distances**. www.thegreenwichmeridian.org/tgm/articles.php?article=9

53 **Ole Rømer and the speed of light measurement**. www.speed-light.info/measure/roemer.htm

54 **Hohmann transfer orbits for space probes**. www2.jpl.nasa.gov/basics/bsf4-1.php

55 **The discovery of Neptune**. en.wikipedia.org/wiki/Discovery of Neptune

56 **Wien's Law linking colour to temperature**. http://mwmw.gsfc.nasa.gov/mmw_bbody.html#wien

57 **The Rutherford-Bohr Model of the Atom**. en.wikipedia.org/wiki/Bohr_model

58 **The structure of the Milky Way Galaxy**. http://solarsystem.nasa.gov/multimedia/display.cfm?IM_ID=8083

59 **Hubble's classification of galaxies**. en.wikipedia.org/wiki/Hubble_sequence

60 **Hubble's Law**. http://imagine.gsfc.nasa.gov/YBA/M31-velocity/hubble-intro.html

10 Pieces of Exciting Astronomy Software

61 **Stellarium** is a free open-source planetarium program, which shows a realistic 3-D view of the sky. www.stellarium.org

62 **Sky-Map.org** is an online viewer using your browser. www.sky-map.org

63 **Cartes du Ciel** (or skychart) is a very easy to use planetarium program. www.ap-i.net/skychart/en/start

64 **Home Planet** is free software that excels at locating satellites, comets and asteroids. www.fourmilab.ch/homeplanet

65 **Celestia** is a space travel simulator that allows you to voyage through the Solar System, to some of the stars in our galaxy and beyond. celestia.en.softonic.com

66 **An Online Orrery** shows you the positions of the planets as seen from outer space for a wide variety of dates. An orrery is a mechanical model of the Solar System, an early example of which was made by George Graham for the Earl of Orrery. http://in-the-sky.org/solarsystem.php

67 **Google Sky** allows you to view celestial objects, including the Moon, planets, stars, our galaxy and other galaxies. www.google.com/sky

68 **Starry Night** is a user-friendly software package with many advanced features. astronomy.starrynight.com

69 **Voyager 4.5** Carina Software's premium package allows you to view the wonders of space. www.carinasoft.com

70 **Redshift 7 Premium** is a great guide to make your home astronomy ventures more thrilling, fun and educational. www.redshift-live.com/en/shop/products/23228.html

10 Mind-Bending Astronomical Distances

71 The **Moon** is some 380,000 km (240,000 miles) from Earth.

72 The **Sun** is around 150,000,000 km (93,000,000 miles) from Earth, also known as one astronomical unit (AU).

73 **Mercury**, the closest planet to the Sun, is 0.387 AU away from the centre of the Solar System.

74 **Jupiter's orbit** from the Sun varies between 4 .95 and 5.46 AU.

75 **Neptune's distance** from the Sun varies between 29.76 and 30.36 AU.

76 The **Kuiper Belt** stretches from 35 to 1,000 AU.

77 **Sedna** – a dwarf planet – is the most distant object in our Solar System. It is 8 thousand million miles from the sun, or 86 AU.

78 The **most distant star** is 55 million light years away.

79 The **most distant galaxy**, in the local group of galaxies, is 3 million light years from us.

80 The **most distant object** in the universe is 30,000 million light years away.

10 Uses of Artificial Satellites

81 A new satellite, the **Global Precipitation Measurement Satellite**, forms an international mission that will set new standards for measuring precipitation, providing the next generation of observations of rain and snow worldwide every three hours. The mission will provide data that will advance our understanding of water and energy cycles, and extend the use of precipitation data, which will directly benefit society. It is hoped that it will help us to understand the severe weather conditions we experience worldwide.

82 **Navigation satellites**. The global positioning system or GPS has virtually replaced the earlier satnav systems, and it can be used by all kinds of travellers.

83 **Communication satellites**. These are used to transmit data, voice and television programmes all over the world.

84 **Mapmaking from satellite**. The taking of detailed photographs from space has allowed cartographers to make improved maps of the Earth.

85 **Weather forecasting**. By studying the movements of cloud formations on a global scale, meteorologists have been able to make better forecasts.

86 **Infrared detectors on satellites**. Although there are infrared telescopes on mountaintop locations, these are still hampered by the atmosphere. Satellite detectors give us much more information.

87 **X-ray astronomy satellites**. Many celestial objects, including our Sun and supernovae, as well as the remnants of supernovae, emit X-rays, but these rays interact with the atoms and molecules of the atmosphere long before they reach the Earth. The only way to study the X-rays from celestial objects is to use instruments on board satellites.

88 **The Hubble Space Telescope** took optical astronomy above the constraints of our terrestrial atmosphere and has yielded some of the best and sharpest photographs of a wide variety of cosmic objects.

89 **Gamma-ray telescopes**. These are very similar to X-rays, but measure shorter wavelengths. They have been useful in pinpointing the sources of cosmic ray particles in our galaxy and elsewhere.

90 **The Planck Observatory**. This observatory was launched to study cosmic background radiation. It has produced detailed maps of some of the earliest radiation in the universe.

10 Places of Astronomical Interest

Some of the best places to learn about the heavens are listed below:

91 **The Royal Observatory, Greenwich**. www.rmg.co.uk/royal-observatory

92 **Institute of Astronomy at Cambridge University**. www.ast.cam.ac.uk/public

93 **Jodrell Bank Discovery Centre**. www.jodrellbank.net

94 **The University of Hertfordshire** has an observatory at Bayfordbury. Bayfordbury.herts.ac.uk/about-the-observatory.htm

95 **The University of London Observatory (ULO) at Mill Hill.** www.ulo.ucl.ac.uk

96 **School of Physics and Astronomy at Cardiff University.** www.astro.cardiff.ac.uk

97 **Keele University Space Observatory.** www.astro.keele.ac.uk/Observatory

98 **Royal Observatory Edinburgh Visitor Centre.** www.roe.ac.uk/vc

99 **Armagh Observatory and Planetarium**, in Northern Ireland. www.armaghplanet.com

100 **The Observatory Science Centre at Herstmonceux.** For many years this was the home of the Royal Greenwich Observatory, when it moved its research facilities to the clearer skies of Sussex. www.the-observatory.org/openevenings

Notes

1 From the *Rubaiyat of Omar Khayyam* verse 52 (Translated by Edward FitzGerald, 1859, London).

2 From a speech given at the Joint Session of Congress on 'Urgent National Needs' 25 May 1961 by US President John F. Kennedy http://www.jfklibrary.org/JFK/JFK-Legacy/NASA-Moon-Landing.aspx.

3 Professor Zdeněk Kopal (1986) *Of Stars and Men: Reminiscences of an Astronomer* (London: Taylor & Francis).

4 Neil Armstrong's message as he first walked on the lunar surface http://www.nasa.gov/mission_pages/apollo/apollo11_40th.html.

5 Christopher Riley interview in *The Observer* 16 December 2012 http://www.theguardian.com/science/2012/dec/16/apollo-legacy-moon-space-riley.

6 A speech made by Sir John Herschel to the British Association for the Advancement of Science in Southampton on 10 September, 1846.

7 William Herschel in a paper to the Royal Society on 26 April 1781.

8 William Herschel in 1783 in a letter to the President of the Royal Society, Joseph Banks. Collected in 'The Scientific Papers of Sir William Herschel', pp. 100–101.

9 The details of the correspondence between Le Verrier and Galle http://www-groups.dcs.st-and.ac.uk/history/HistTopics/Neptune_and_Pluto.html

10 Ibid.

11 Professor I. Bernard Cohen 'Towards Newtonian Gravitation' delivered at a conference to celebrate the 300th anniversary of the founding of the Royal Observatory at Greenwich.

12 Albert Einstein quoted in an article entitled 'Time, Space and Gravitation: The Newtonian System' *The Times* 28 November 1919.

13 Brian Cox (2010) *The Wonders of the Solar System* DVD and associated 2010 BBC book title (London: Collins).

14 Jacob Bronowski (2012), *The Ascent of Man* (first pub 1973) (London: The Folio Society).

15 Sophie Curtis 'Colossus at 70: the computer that helped bring down Hitler' interview in *The Daily Telegraph* 5 February 2014 http://www.telegraph.co.uk/technology/news/10618866/Colossus-at-70-the-computer-that-helped-bring-down-Hitler.html.

16 Ibid. Quote from Tim Reynolds, Chair of the National Museum of Computing.

17 Thomas Hardy, *Far From the Madding Crowd* (1874) Chapter 2.

18 Professor A. H. Cook (1973) 'The Astronomer as Natural Philosopher'. His inaugural address as Jacksonian Chair at Cambridge University. (Cambridge: Cambridge University Press).

19 Sir Arthur Stanley Eddington (1926) foreword to *The Internal Constitution of the Stars* (Cambridge: Cambridge University Press).

20 A. Hallam (1989) *Great Geological Controversies* (Oxford: Oxford University Press).

21 Ibid

22 Sir Arthur Stanley Eddington (1926) foreword to *The Internal Constitution of the Stars* (Cambridge: Cambridge University Press).

23 Galileo (1610) *Sidereus Nuncius* (The Sidereal Messenger). Reference found in (1973) Stanley L. Jaki *The Milky Way: An Elusive Road For Science* (Newton Abbot: David & Charles).

24 Georges Lemaître (1946) *L'Hypothèse de l'atome primitif (The Primeval Atom* Translated by Betty H. and Serge A. Koff, New York, van Nostrand, 1950).

25 ESA's (European Space Agency's) Planck Telescope team 2013 sci.esa.int/planck/53108-planck-and-the-cosmic-microwave-background/.

26 Edwin P Hubble (1946) 'The Exploration of Space' *Popular Astronomy* vol 54 p.183.

Index

Acknowledgments

Picture credits: Chapter 1 Buzz Aldrin – NASA; Chapter 2 Solar System – Aaron Rutten/Shutterstock.com; Chapter 3 Planetary probe – previously appeared p299 *Das Ticken des Kosmos* (Elsevier GmbH) by Percy Seymour and Dennis Bacon; Chapter 7 Milky Way – NASA JPL; Chapter 8 Hubble's Classification of Galaxies – NASA.